BY THE SAME AUTHOR OF:

GOLD MINING IN THE 1980's – THE COMPLETE BOOK OF MODERN PROCEDURE

ADVANCED DREDGING TECHNIQUES – PROFESSIONAL GOLD DREDGER'S HANDBOOK

ADVANCED DREDGING TECHNIQUES VOLUME 2 – PART 1 FINDING AND RECOVERING PAYSTREAKS

ADVANCED DREDGING TECHNIQUES VOLUME 2 – PART 2 SUCCEEDING AT A GOLD DREDGING VENTURE

ADVANCED DREDGING TECHNIQUES
VOLUME 2 PART 1

TABLE OF CONTENTS

ADVANCED DREDGING TECHNIQUES

VOLUME 2 – PART 1

FINDING AND RECOVERING PAYSTREAKS

By:

DAVE McCRACKEN

FRONT COVER: Dredger using his new "Triple Sluice dredge" photo by jerry Wayne Keene

All other photos throughout this book were taken by Roderrick L. Long Original drawings done by Barbara Long. Final art work by Don A. Corrie

First Printing December 1981
Revised in February 1983 , May 1986

PRINTED AND BOUND IN THE THE UNITED STATES OF AMERICA

PUBLISHED BY KEENE INDUSTRIES
9330 Corbin Avenue
Northridge, California 91324
(818) 993-0411

SECOND PRINTING 1997

PART 2

ACKNOWLEDGEMENT

It would simply take too many words to properly express my thanks to Harlan Cockroft (Red) for his help during my early days of gold dredging. My sucess as a professional gold dredger -- and therefore the knowledge contained in these books on the subject -- is largely due to Red's unending willingness to give a hand.

To: Anita - my agent,
assistant, backer, banker, proof reader,
advisor, helper, and loyal friend.

IMPORTANT NOTE: In Studying this manual, be very very careful not to go past any word which you do not fully understand. Recent breakthroughs in the field of study have shown that words are symbols which relate concepts. To not understand or to misunderstand a word, is to not understand or misunderstand a concept, which can thereafter make the subject confusing to you. It is a good idea to have a good English dictionary handy when studying this or any other subject. I have also provided a glossary of mining terms at the back of this manual for your easy reference. For further information regarding the mentioned breakthroughs in the field of study, write to:

Bridge Publications
4833 Fountain Avenue
East Annex, Los Angeles, CA 90029

INTRODUCTION

There is a certain amount of social and economic freedom that goes along with being a successful gold dredger. You are recovering for yourself a substance–gold–which is internationally acceptable as an exchange medium, and which maintains or increases its exchange value over time–despite what the other currencies and exchange mediums are doing. Gold is going to be valuable–no matter what happens with the economic situation; but if things do continue to get worse, gold is just going to keep on becoming more and more valuable. So there is a great deal of security in having gold–or in being able to recover it in paying quantities.

Presently, the majority of our most prominent economic authorities are saying with assurance that we are going to see gold trading at the two to three thousand dollars per ounce levels–and higher–within just the next few years! They also say that gold is presently on that long upward trend to reach these new high peaks–before dropping down to a new low in the eight to nine hundred dollars per ounce range. These experts are not guessing, hoping, or kidding around about this; they are financial investment advisors, whose bread and butter depends on their accuracy and success at predicting what the various commodities–like gold–are going to do. Some of them have never been wrong yet in any single prediction about what the gold market was going to do, and they are willing to stake their reputations upon their CERTAINTIES that the gold prices are going up to these astronomical levels. They have very solid economic reasons for making these predictions.

One of the main economic reasons why we can expect to see the gold prices continue on a long range upward trend is because inflation is continuing—and it is most likely going to get worse as time moves forward. Any increase in the money supply is a direct cause of more inflation. When the government spends more money than it has on hand (deficit spending), it prints up new money to make up the difference. This new money will eventually make all the existing money that is already in circulation worth just that much less in its trading value. And so you get inflation; the dollar isn't worth as much; so it takes more of them to buy the same amount of goods and services.

As inflation continues to make the dollar worth less and less, the gold prices will continue to go up and up. This is inevitable; because if the dollar is less valuable, it will take more of them to buy anything that is valuable—like gold. But there are other reasons as well: As more and more people realize that the dollar is continuing to lose its trading value over time, and that gold holds or increases its trading value over time, we will see more people taking their assets out of the dollar and putting them into gold. Gold—just like any other commodity—is going to react to supply and demand factors. As the demand for gold increases, the prices will go up. So as the dollar continues to inflate, we can expect to see the demand for gold increase, which will drive the prices up, and so on.

During the period of time between 1975 through 1978, the combined deficit spending for the U.S. government was 219 billion dollars. Inflation follows just behind an increased money supply—and the gold prices follow inflation. Gold prices peaked at $850/ounce in 1980. In 1982 the U.S. government overspent $230 billion—that's well over three times as much as the previous period mentioned above! The White House says the 1983 deficit will reach at least $208 billion, that 1984's deficit will reach $189 billion, and 1985's deficit will reach $194 billion. Even with all the effort and noise we hear about cut-backs, we are still talking about an incredible amount of government overspending. Presently, the money supply is expanding at an annual rate of 16%!! This is going to be the direct cause of further inflation—lots of it; it just takes awhile to catch up. And this will cause the gold prices to continue on upward as well. There appears to be no other way thay can go on the long trends until the money presses are shut down—and there are no plans at all to shut the money presses down.

This all adds up to a very bright future for the gold prices; and, I'm sorry to say, a rather poor future for the dollar and the economy—and for those prople who are fully dependent on them.

And it also all adds up to some very good reasons why it is worthwhile to be involved with gold dredging—and why it is a good idea to be really good at it. With the prospects of future gold prices being what they are, there are very good possibilities that the successful gold dredger will be able to make a fortune at it. And in view of the strong possibility of worsening economic conditions as time moves forward, if nothing else, the successful gold dredger will have something worthwhile to fall back on if the need arises.

There is no doubt about it, with conditions being what they are, gold dredging—either as a hobby or a professional activity—carries a bright future along with it. Gold dredging, as a rewarding and beneficial activity, has never looked more positive than it does today.

Thank you, by the way, for sending along all those great letters. I really do appreciate hearing from you guys and gals out there, and I always make sure to answer each and every letter I receive. This has kept me pretty busy at times! Your communications and encouragements resulting from my earlier books on mining are what have prompted me to write this final set of volumes on dredging.

I say "final" because at this point, dredging for gold has been pretty much developed into an exact technology—with exact procedures. All of this—or at least the vast majority of it—is contained in this two part volume, and in my two earlier books on the subject; and right now I really don't see why any additional books on the specific subject of gold dredging would be necessary.

We have learned quite a lot since the first volume of **ADVANCED DREDGING TECHNIQUES** was written. All of the information in that book is still valid, however, and vital to know. I haven't repeated it here; this is mostly new and additional information which adds to what has already been written in my earlier two books.

One of the most important developments in modern gold dredging is that we now know precisely what paystreaks are; and we know exactly why, when, and where they form in a running body of water. We also know how and where to look for them; and we have refined our sampling techniques to the point where they will lead us right into the paystreaks everytime—if we just follow through and do them standardly. We know quite a lot about these things at this point, and I have done my best to communicate the whole works to you in these books.

Actually, though, while there is a great deal of information, with much of it being quite vital to know if a dredger wants to succeed continuously, most of it is common sense by nature; and the subject is not at all hard to understand once you start getting the hang of it. One thing to know—and remember—is that the procedure of getting into paystreaks is now very exact. If you do this, this, and this, you will get into gold—and it's as simple as that. We are working with very precise and workable procedures. We have taken all reliance on luck, hope, fate, superstition, and all the rest, and we have put it on the shelf. In turn we have replaced it with observable truths, workable procedures, common sense, lots of hard work, and reliance on your own abilities, and have given you the "sure thing" methods of putting yourself into paying quantities of gold on a continuous basis.

But knowing this information is only about a third of it. You still have to be able to apply it—and you still have to go out and DO IT. So you are also going to have to do your part, and that's going to require plenty of hard work. However, if you can understand, apply, and carry out the principles laid out in these books, you will bring in a much better return for your hard work than you would otherwise; you will get into paying quantities of gold on a consistent basis--there is simply no doubt about it. And your success will have nothing at all to do with luck; you will be creating your own fortunes. At this point, our knowledge and our procedures are 100% workable—and I'm proud to be able to present them to you. Now, it remains to be seen how good you can put it to use! I'm not going to wish you luck, because at this point, you shouldn't need it any more!

Best regards,
Dave Mac

Chapter 1

PAYSTREAKS–WHAT THEY ARE

MAJOR STORMS

It only takes a small amount of water force to move sand sized particles downstream in a riverbed. It takes more water force to move pea sized gravel down river, and more force than that to move golf ball sized rocks, and even more to move baseball sized rocks, and so on.

It also takes a greater amount of water force to move larger sized particles of gold than it does to move smaller ones–just like the other streambed materials. There is a massive amount of gold in gold bearing waterways which is so small in particle size, that it floats in a state of suspension in the river water itself. Some amount of gold is moving downstream in any gold bearing river at any given time. An increase in water flow will also increase the amount of streambed material and gold being moved downstream.

For the most part, the normal amount of gold being moved downstream in a river is insignificant to a gold dredger; because it is so minute in size, that it defies normal recovery methods; and is so widely dispersed in small amounts throughout the river, that it simply is not worth going after in itself with the standard type of suction gold dredge.

Normal winter storms, for the most part, move incredible amounts of fine and ultra fine gold down the gold bearing rivers. But this gold is likely to be so widely dispersed throughout the overall river, that it is of no great value to the gold dredger either.

The gold dredger is mostly interested in that gold which lies inside and underneath of formed streambeds. For the most part, this gold will stay right where it is until a storm of major proportions comes along and causes so much water force in the river, that large sections of formed streambeds—and the gold that is with them—are swept up and washed downstream.

When a storm or snow runoff of huge dimensions comes along and creates so much force of water flow that large portions of the riverbed are torn up and washed down along the bedrock, large portions of the bedrock will also get pounded loose; and any gold which was trapped in that bedrock will also get washed further downstream along with the rest of the streambed material.

The amount of gold which is still sitting inside of streambeds in the proven gold bearing rivers is incalculatable—there is a whole lot of it! Much of it was left behind in low grade deposits which the earlytimers were not interested in mining; because, while there might have been a lot of gold in some sections of river, it might also have been too widely dispersed or sitting underneath too much overburden to make it worth mining in those days. Other very rich deposits were entirely missed because they were out of sight, and without processing every bit of streambed—which was overwhelmingly impossible—they simply could not find them all. A lot of gold was never recovered, and was washed out of the high bench deposits, through sluice boxes, and right back into the present rivers and creeks—as in the case of hydraulic mining, where an estimated 50% of the gold was missed by many of the large and small operations alike. Also, the last 100 years or so of natural and man caused erosion has washed a great deal more gold into the present rivers and streams out of the higher and older dry streambeds–and perhaps even out of some lodes that are still in existence–although this is probably not as large a contributing factor as these others are in most cases. In taking all of this into consideration, we are talking about a lot of gold still being in these gold bearing rivers–in many cases, more than has already been mined out of them.

When a major storm occurs in a gold bearing area and tears up large portions of streambed, a great deal of gold is set free and put into motion downriver. A fair amount of this newly released gold, because if its superior weight, will be deposited in common areas along the riverbed. This is the type of gold that the modern gold dredger is most interested in. The same major storm which causes enough force to tear up large portions of streambed material, will also redeposit most of that material into newly formed natural streambeds–even in those same areas which were once mined by the earlytimers.

For the most part, the normal winter storms occurring in gold country do not set up enough water force to do this. A winter storm might be enough to sweep up small portions of streambed in faster moving sections of river and redeposit new streambeds in those areas; but this small amount of movement is not likely to put paying quantities of gold into play in the riverbed. It takes a HUGE amount of streambed in motion to place HUGE amounts of gold into movement. This causes a whole river to form new placer deposits. Such storms occur occasionally, and are the main cause for a streambed cutting deeper into the earth as time goes along. Most gold bearing areas have had at least one or two of these major storms since the early 1960's. Alaska has major storms along with massive snow runoffs, and flood forces like this happen more often.

One way to tell if a storm is actively moving a lot of the streambed, is by listening to the noise that the river is making. When millions and millions of tons of rocks, cobbles, and boulders are being swept downstream along the bedrock foundation during a huge storm, the ground shakes and vibrates; and the river rumbles like a huge loaded freight train. After a major storm has been through an area, the plant growth, underbrush, and weeds which normally grow on the river gravel bars will be gone–washed out. This will also be true with much of the growth–and even many of the trees–along the riverbanks.

It takes an incredible amount of water force to cause an entire riverbed to move downstream, but this is what it takes to form many new placer redeposits in the river. The earlier idea that gold drops into the river and simply is vibrated down through already formed streambeds, to eventually reach bedrock and form a placer deposit, is very limited in it's workability and will not lead one to find where the paying deposits are usually found. An understanding of the above information, and what is to follow,will give you a basic foundation for learning how to find and recover the different kinds of streambed deposits.

DIFFERENT KINDS OF STREAMBED PLACER DEPOSITS

There are several different kinds of streambed gold deposits that a gold dredger needs to be aware of, because each needs to be searched for and mined in a different way.

TAILRACE AND HIGH BENCH BLEED-OFFS: A high percentage of the gold in today's rivers was once washed into them out of the higher and older streambeds resting up alongside the rivers, or those which are located some distance away. This is the result of about two million years of massive flood storms and natural erosion. Today's storms continue to wash gold out of the banks and into the present rivers and creeks. That gold which was set free by earlier mining activities, but missed by their recovery methods, is likely to be washed into the present streams and creeks too.

Because gold tends to follow and deposit along a certain common path in the river, when it is entered into the river from a point off this common path, it will have to be washed some distance downstream before it finally arrives in the common path. Therefore, these new deposits of gold from high bench bleed-offs and tailraces are likely to follow a path of their own downriver until they get into the main path. So you would need to look for this type of deposit usually in a different section of the river than you would normally look. (See figure 1-1).

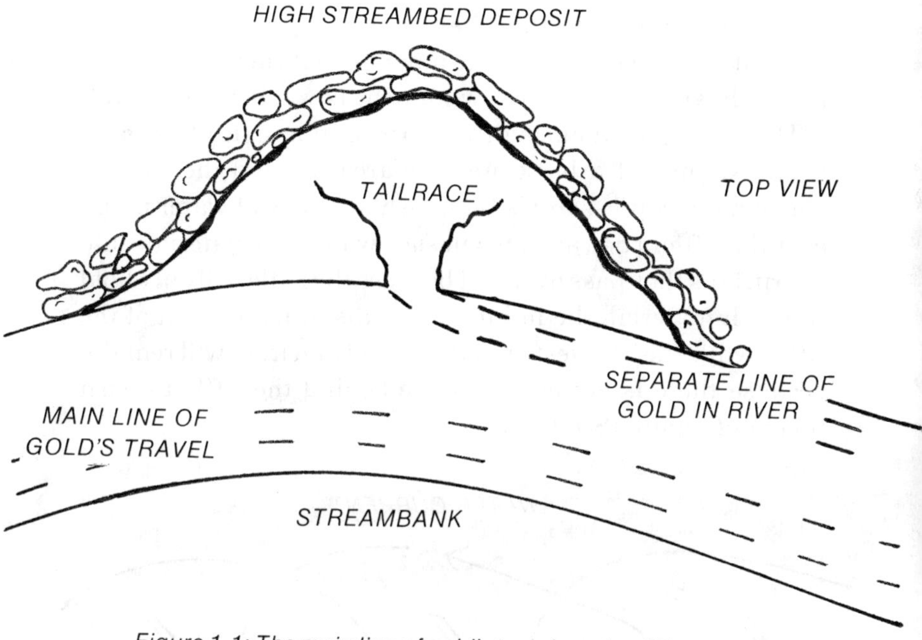

Figure 1-1: The main line of gold's travel can be different than the gold line from a high bench or tailrace bleedoff.

PAYSTREAKS: Paystreaks -- usually called "Stringers" by gold dredgers -- are the most common, and best paying, and easiest to generally locate, type of placer gold deposit being found in most of today's gold bearing rivers. A "Stringer" is a concentration of gold which has formed in a section of river because of a sudden or gradual lessening of water force in that area during a major storm. During an extreme highwater period, when the entire streambed is moving along the bottom of the river, because there are many sudden and gradual changes in the river's direction and downward slope, different areas along the bottom of the river will have greater or lesser forces than others, and different push/pulls are set up along the river's course. There will be high pressure areas, medium pressure areas, low pressure areas, and even back pressure areas. When you see water moving backwards against the main flow of water--as in a back eddy -- you are looking at a back pressure area. Where you see the water slow down or even stop, in comparison to the main flow of water, you are looking at a low pressure area. In a river, during a massive storm, many low pressure and back pressure areas are created along the river's path. These are the places that gold concentrates inside of in large amounts -- if they are lying in the path which gold is following as it is washed downriver.

The standard hungarian type riffle in a sluice box is an ideal example of a back pressure area. Behind the riffle, a vacuum (back pressure) is set up when a flow of water is run over the riffle. The vacuum will suck water and materials into the riffle as they pass over; but because the riffle will only hold so much material, the lighter materials are forced out of the riffle, while the heaviest materials (gold and iron) will remain; because there is not enough force behind the riffle to push them out again. (See figure 1-2).

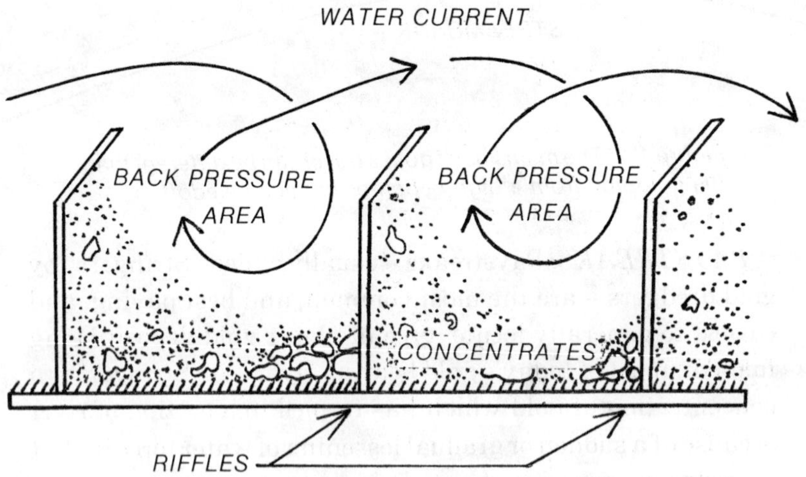

Figure 1-2: A vacuum is created just behind a hungarian riffle when water is passed over it.

Similar conditions to this will occur in a streambed in many places during a major storm, setting up back pressure areas like this and on a much larger scale. Many of the common types of low/back pressure areas that are found in a river are shown in Chapter 4 of Volume 1 of this manual.*

Low and back pressure areas can also be formed because of some major change off to one side of the river, or when the river makes a change of direction. One good example of this is the common low/back pressure area that will be found at the tail end of most bends which a river takes. If you take a good look at where a river changes direction, you will see

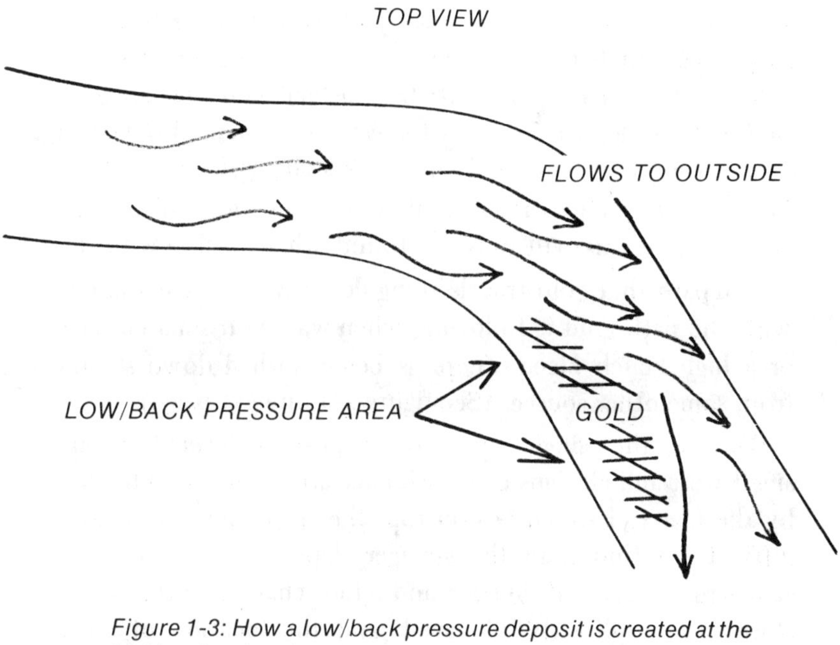

Figure 1-3: How a low/back pressure deposit is created at the inside of a river bend.

**ADVANCED DREDGING TECHNIQUES VOL. 1 is available by writing KEENE INDUSTRIES, 9330 Corbin Ave., Northridge, CA 91324.*

that the flow of water and material will come on in full force at the head of the bend. However, since centrifugal force will cause a great deal of the pressure to flow to the outside of the bend, there will be a lowering of pressure—or even a reverse pressure (back pressure)—to the inside of the rear-end of the bend. And sure enough, the inside tail ends of river bends are a very common place to find a sizeable stringer (paystreak) in a gold bearing river. So stringers can be caused by a change in pressures off to one side of the river, or by a change in pressures due to a difference in downward slope of the river's foundation. (See figure 1-3 on previous page.)

THE SINGLE DEPOSIT: A single deposit of gold can form anywhere along the general path in the river that gold follows, where an obstruction, a small change in bedrock, or a crevice is located. In this case, you are not going to find a large area which has been concentrated with gold—like you will withh a stringer deposit; but instead you are going to find a single deposit inside of the crevice, or behind the small obstruction, or whatever. Sometimes a single deposit will be found that carries a substantial amount of gold. Single deposits like this will only be found when in line with the general path that gold travels along downriver, or when in line with the path gold is following when washed from a tailrace, or a high bench bleed-off, or is being washed down stream from some other source. (See figure 1-4 on next page.)

Because the single type of deposit is defined to one small area, and all signs of its presence are usually well hidden by the material which lies on top, the single deposit is more difficult to find than the stringer deposit. In the case of stringers, many of the major and minor changes in the river's flow can be directly observed, and the location of their resulting low and back pressure areas can be easily estimated. Stringer deposits cover some width and distance, and so are much easier to locate by sampling—and because stringers usually carry a sizeable amount of gold, you'll find that most experienced gold dredgers are going after them, instead the single type deposits.

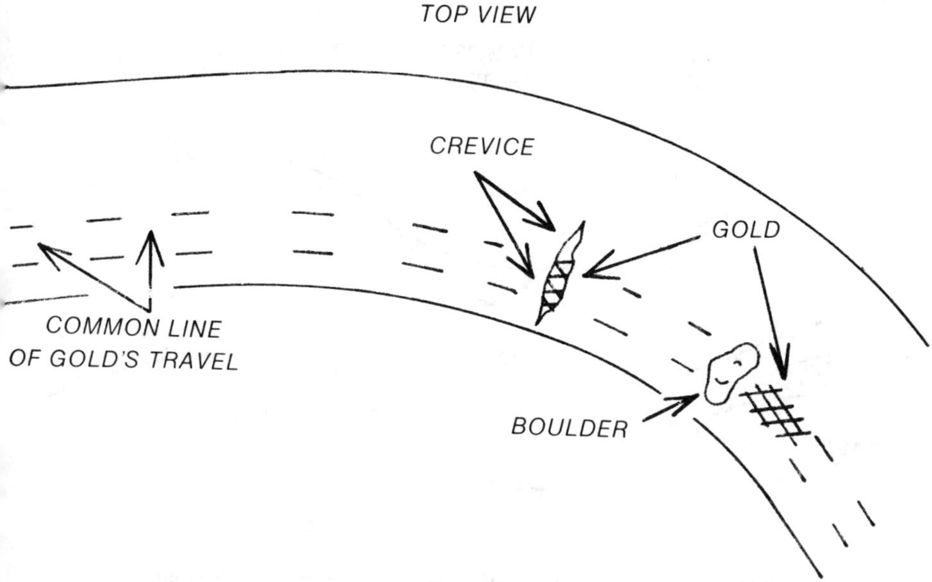

Figure 1-4: Single gold deposits will be found along the common line of gold's travel.

THE GENERAL PATH

Because gold is six or seven times heavier than the average of other streambed materials that get swept downriver in a major storm; it tends to follow and be redeposited along a common path. For the most part, this path seems to follow to the inside of the bends, and the shortest route between the bends, as shown in figure 1-5. (Next page)

Because during a major storm, there are so many different forces at work in a riverbed, and there is so much material being tumbled downstream, it's reasonable to assume that a lot of the gold—maybe even most of the gold—doesn't travel down along this path in all cases. However, the statistics of history do show that a far far greater percentage of single deposits and stringers have been found along this common

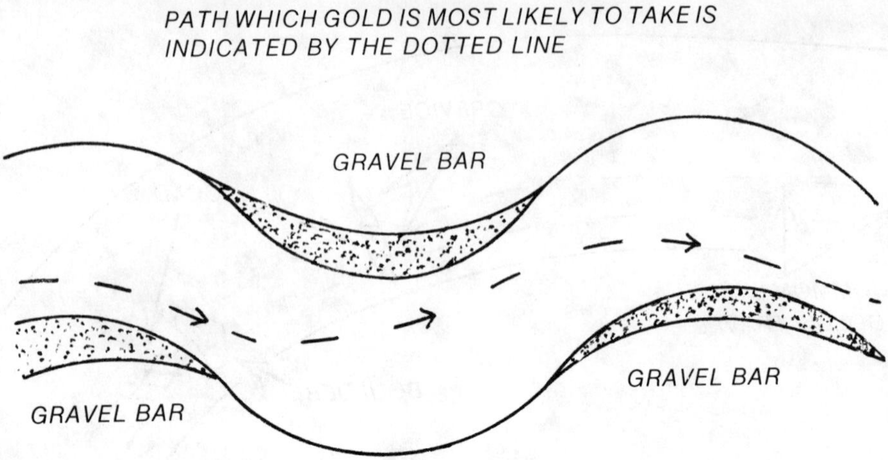

Figure 1-5: The general path which gold tends to deposit along.

path, than have been found elsewhere in the river. So it's worthwhile for the gold dredger to know about the general path, and to direct his sampling activities towards it. For now on, I'll refer to this path as the "general line of gold's travel."

The general line of gold's travel can be largely affected by the various major low/back pressure changes in the stream's flow—especially those that tend to pull from off to one side of the river. The general line is usually most definite just after a major bend in the river, where the huge low/back pressure to the inside tends to pull the line strongly over to that side. On a long stretch of straight river, where there are no major low/back pressure areas off to either side, the minor ones tend to disperse the general line of gold out into a less defined area of travel, as shown in figure 1-6.

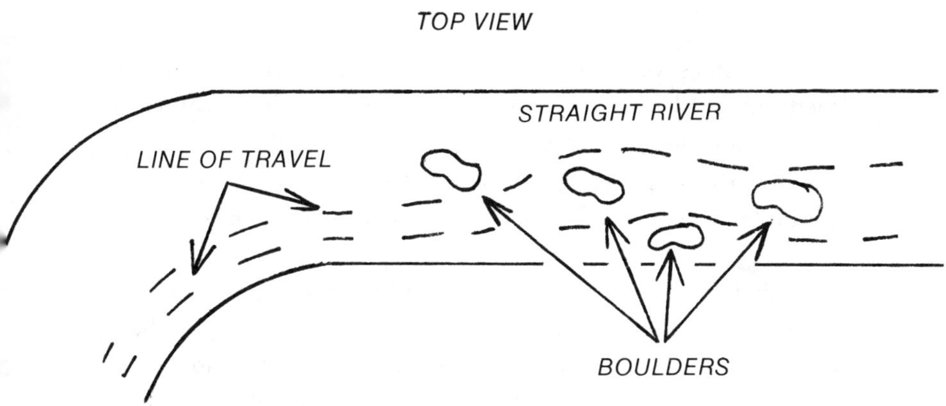

Figure 1-6: The general line of gold's travel can cover a less defined area after a straight stretch of river.

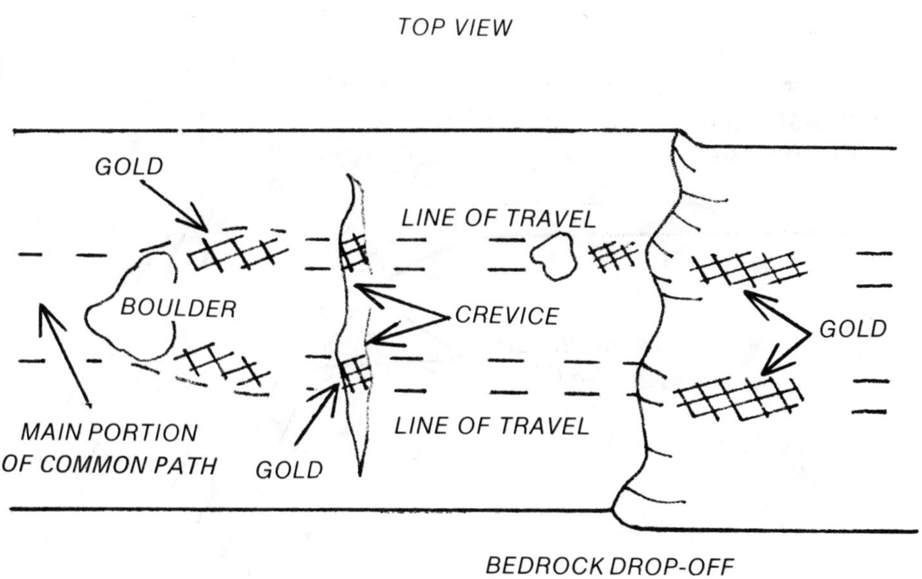

Figure 1-7: More than one general line of gold travel can form on a straight stretch of river.

Sometimes on a long stretch of straight river, small and medium sized low/back pressures, or obstructions, can cause more than one general line of gold to form and travel down the length of river; so single deposits and stringers can be found and recovered along each separate line, as shown in figure 1-7. (Previous page)

Sometimes there can also be more than one general line of travel down a river even when there are strong low/back pressure created off to one side—like a major bend. These additional lines are usually the result of a tailings or high bench deposit bleed-off from upstream, or a lot of the gold already having been widely dispersed throughout the river because of a long stretch of straight river upstream, as shown in figure 1-8.

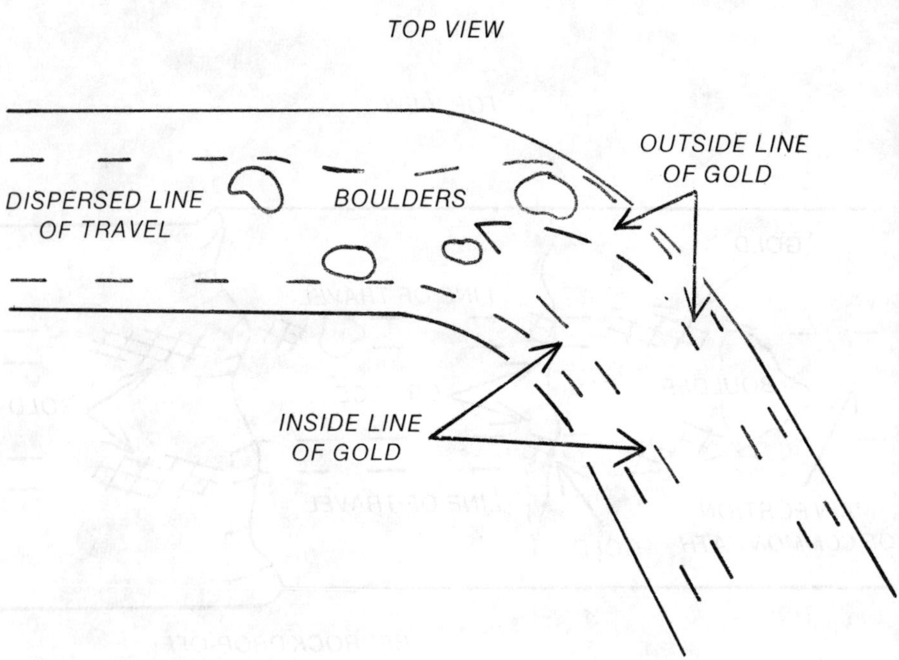

Figure 1-8: Two lines of gold travel due to a large amount of dispersement upstream.

In addition, an increase in the other heavy minerals and materials in a riverbed tends to also concentrate along and follow this same general line that gold travels. Iron rocks, black sands, old nails, pieces of lead, and tools are just some of the heavier materials which commonly follow this same path. However, because iron and lead do not have as high a specific gravity as gold does, the path that these follow can be a wider one and have less defined boundaries than the general line of gold's travel. This is also true of the fine and ultra fine sized particles of gold. The smaller that a particle of gold is, the less it is likely to be affected by the various low/back pressure areas in a riverbed. The greatest concentration of fine gold and other heavy materials, however, does follow the general line or lines of travel that gold normally deposits along.

Those low/back pressure areas which are not on gold's line of travel may show an increase in the amount of other heavy materials, and possibly even a small amount of fine flood gold; but you are not likely to find paying quantities of redeposited gold in them. The low/back pressure areas which are positioned along the line of gold's general path will show the greatest potential of paying gold deposits, and are where you're likely to find the greatest amount of other heavy materials too. Sometimes you can find the common gold path by sampling around to find where in the river that the greatest amounts of heavy materials are concentrating. From there, it is possible to sample forward along that line to find where the low/back pressure areas (stringers) are located. If there is an increase in the heavier minerals and materials showing, and if they are along the common gold path, there should also be at least a slight increase showing of gold along that line too. This is the best sign that you are on the correct line to find the paying deposits. (See figure 1-9).

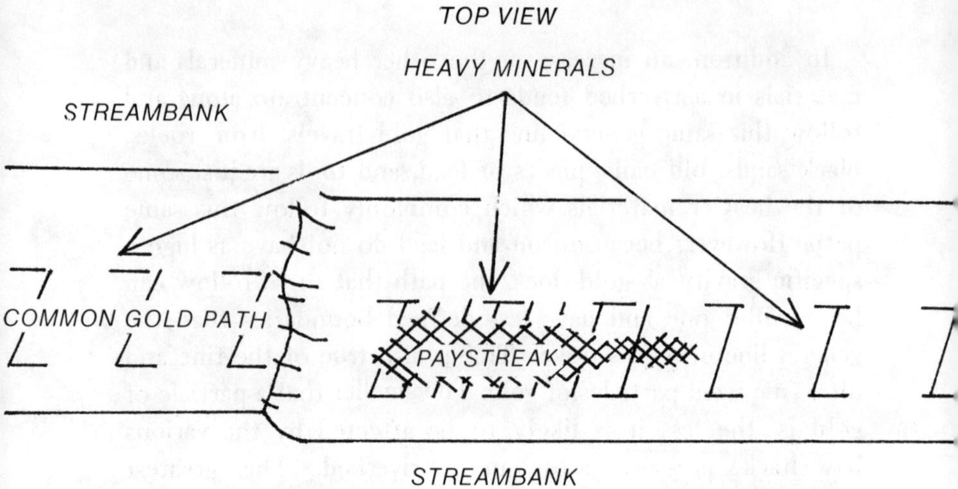

Figure 1-9: An increase in gold and other heavy minerals, and materials in the streambed is the best sign that you are on the common gold path.

Gold does not deposit in paying quantities everywhere along the general line of travel—not by a long shot. Although, you are more likely to find small traces of it having passed through along the line in places here and there where small amounts were able to become trapped (very small single deposits). Very rough and irregular bedrock along the general line will often show traces of gold; whereas smooth and polished bedrock along the general path usually will not.

Once in awhile, a fairly rich single deposit will be found along the general line of travel if conditions are just right for it; but for the most part, it takes a major low/back pressure area along the line of travel to trap gold in paying quantities.

It's very important for a gold dredger to understand what this general line of gold's travel is, and how to find it in a riverbed. Sometimes the general path is hard to find. Sometimes there is more than one path that gold is following. You can always depend on at least one path being present as long as gold has traveled through that section of river—unless it has been thoroughly dispersed by a long straight stretch of river, in which case the entire river may be that path.

Once in awhile, it can be confusing as to what the path is doing—that is, where it's located in the river. It can be pulled off of its apparent course to one side by numerous low/back pressures created down beneath the streambed which cannot be seen at the surface. Some of the time, there might not be any sign of the general path at all—even if you dredge a hole right smack down in the middle of it. This does not matter very much though; because when sampling, you'll be looking for the low/back pressures, and these will always show some sign of the general line of travel if they lie in its path. It just makes things go faster if you can determine exactly where the general line of gold's travel runs in the river. This would be much easier to do if there were no hidden influences at work beneath the streambed. But if that were the case, it would be very easy to find the deposits and they would already be gone!

As it is, up until the present, knowledge of these low/back pressures along the general line of travel has only been known by a privileged few—those who worked at it hard enough, and long enough, to figure it out. And so stringer deposits are still in place in great abundance in the rivers throughout the gold country. These, for the most part, were put in place by the last major storm which occurred in each area (1964 flood in California and Oregon).

THE CHARACTERISTICS OF STRINGERS

In many of the earlier books on gold mining, mention is made that gold will deposit in the black eddies. This is true—to a point. First of all, for large amounts of gold to collect in a black eddy (back pressure area), it will have to be located along the path that gold will follow in its route down river during a major sized storm. And second, if too much force of water is moved over the top of a back pressure area, it can become a boil-out area—in which very little

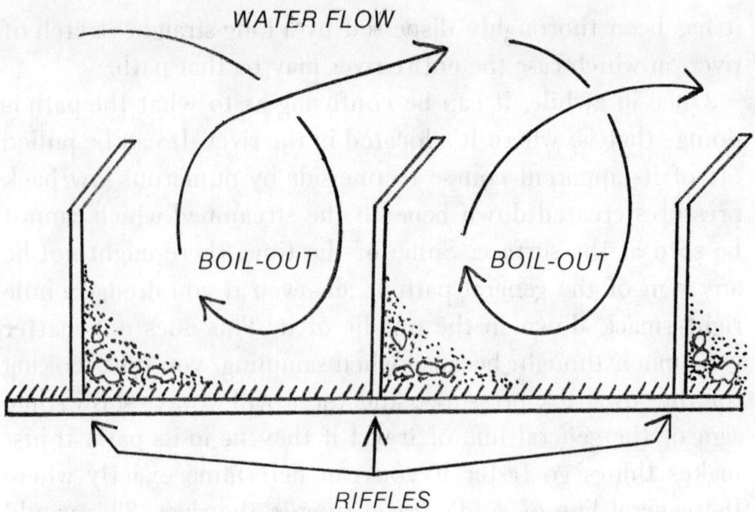

Figure 1-10: Too much water force over a back pressure area can cause a boil-out.

concentration of gold values will take place. An example of this is in the case of a Hungarian-type rifle when too much water velocity is flowed over it—like when the slope of the sluice box is increased too much. When this happens, the force of the flow being pulled into the vacuum behind the riffle becomes so turbulent, that even most of the heavier materials—gold included—are kicked out of the riffle. In fact, the force can be increased to the point where no concentration at all will take place in the back pressure area. (See figure 1-10.) A back pressure area in a river can also become a boil-out if too much water force is passed over it in a storm. This is one of the reasons why you don't find the amount of gold that you expect to find in some prime low/ back pressure areas in a riverbed. It is also the reason why some large boulders do not have paying quantities of gold behind them, even when they are sitting in the center of the gold's main line of travel—and sometimes when sitting right in the middle of a well paying stringer.

I have uncovered quite a few paystreaks which paid quite well everywhere but behind the large boulders. On some of these, I found where some earlier prospectors had come along and dredged sample holes directly behind the large boulders; and since they did not find much gold there, they apparently didn't pursue it further. I could tell by the streambed formation that, other than where the sample holes were dredged, no other mining had been done behind those boulders. The gold simply did not deposit behind them, yet they were sitting right smack in the middle of good paying stringers. (See figure 1-11.)

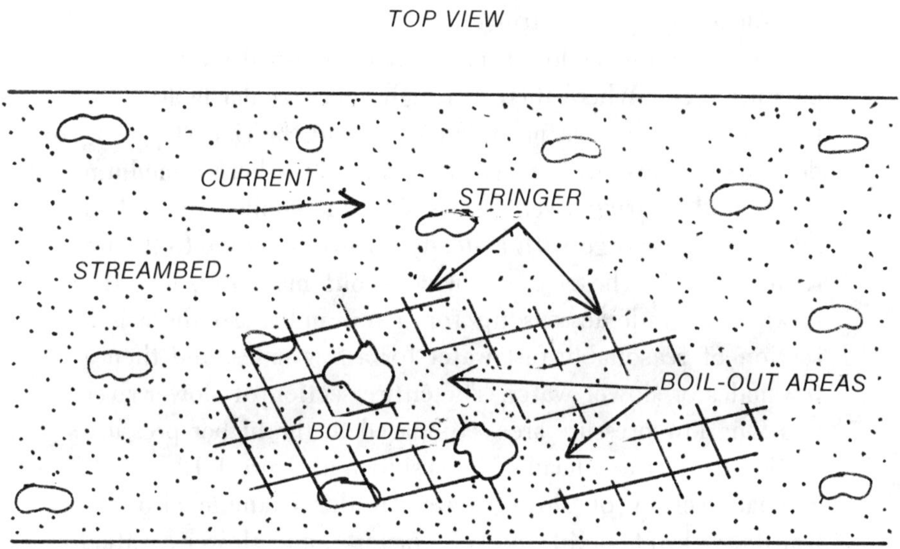

Figure 1-11: Sometimes gold will not concentrate behind the boulders — even when in a stringer deposit.

Had those prospectors moved their holes just a little to the outside of the boulders in either direction, they would have been into gold hot and heavy. This is a good lesson to learn by. In shallow material, I will usually dredge two or three sample holes side by side, and some distance apart, for comparison of streambed material to see where the general path might be located, and to ensure that I do not miss the stringer if it's there.

When we talk about stringers, we're also talking on a much broader scope than just back eddies. A stringer deposit can form wherever the pressure of force, in a major storm, along the line of gold's travel drops off enough to drop gold, but still moves fast enough to sweep off the lighter materials. I have seen a few well paying stringers recovered right off of smooth and polished bedrock; and while this happens occasionally, places where there is a lot of rough and irregular bedrock, in a low/back pressure area, are far more likely to provide a good paying stringer.

There can also be low/back pressure areas within low/back pressure areas. Where there is rough and irregular bedrock in these places, you are most likely to find the richest paying deposits. I know of a place, for example, where a medium sized gold bearing river travels for some miles through a rather narrow gorge with a steady downward grade to it, and then suddenly the riverbed widens out markedly, and the downward grade lessens up for a few miles. So there is a section of reasonably fast water for a few miles, and then a few miles of slower water. The entire section of slower river is a long low pressure area, compared to the higher pressure section of river up ahead of it, as shown in figure 1-12.

That section of slower water in this example also has bends, short sections of faster and slower water, outcroppings, upcroppings, obstructions, major changes in bedrock, and so forth, which set up additional major low/back pressure areas inside of the long length of slower moving water. On top of that, for the most part, the bedrock is very rough and irregular all the way through the entire section of slow moving river. And for its entire length, each & every obvious low/back pressure area has carried a large and well concentrated stringer deposit. There are also several

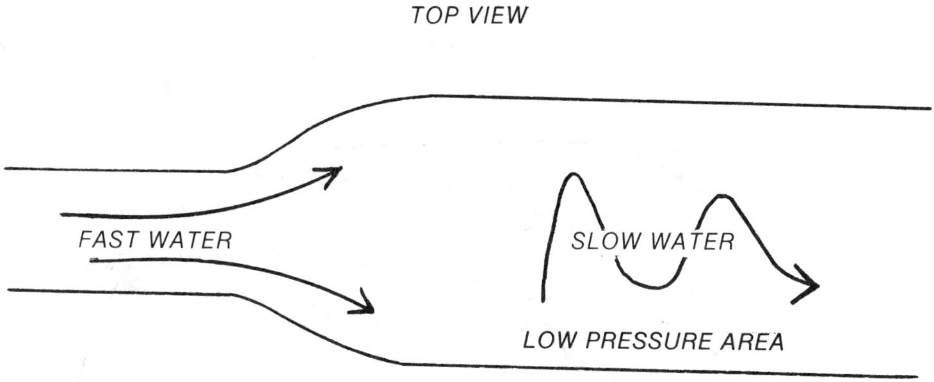

Figure 1-12: Fast water (high pressure) section of river flows into a slow water (low pressure) section.

long straight stretches of river, which resulted in at least three separate and well defined lines of gold in some areas. Because the bedrock was right for it, and because the whole stretch was one huge low pressure area, there were stringer deposits to be found along each line of travel wherever any low/back pressure area was set up in the stream's flow–large or small. Indeed, in many places it was a paying proposition to just get onto any of the gold lines and start dredging; it would pay along the entire line and become very rich in the low/back pressure areas, and where single depsoits could form. My conservative estimate is that well over 2000 ounces of gold have been taken out of a one mile stretch of river in that area in just the past two years alone! One interesting point about this is the entire riverbed throughout that long slow section of river was thoroughly mined during the late 1800's–it paying very well then too! So you can find low/back pressures within larger sized low pressure areas, and this is one thing you'll want to keep your eyes open for.

In some situations, it might take some distance for the water force to slow down enough in a long low pressure area to start dropping gold in large amounts. Like where the river

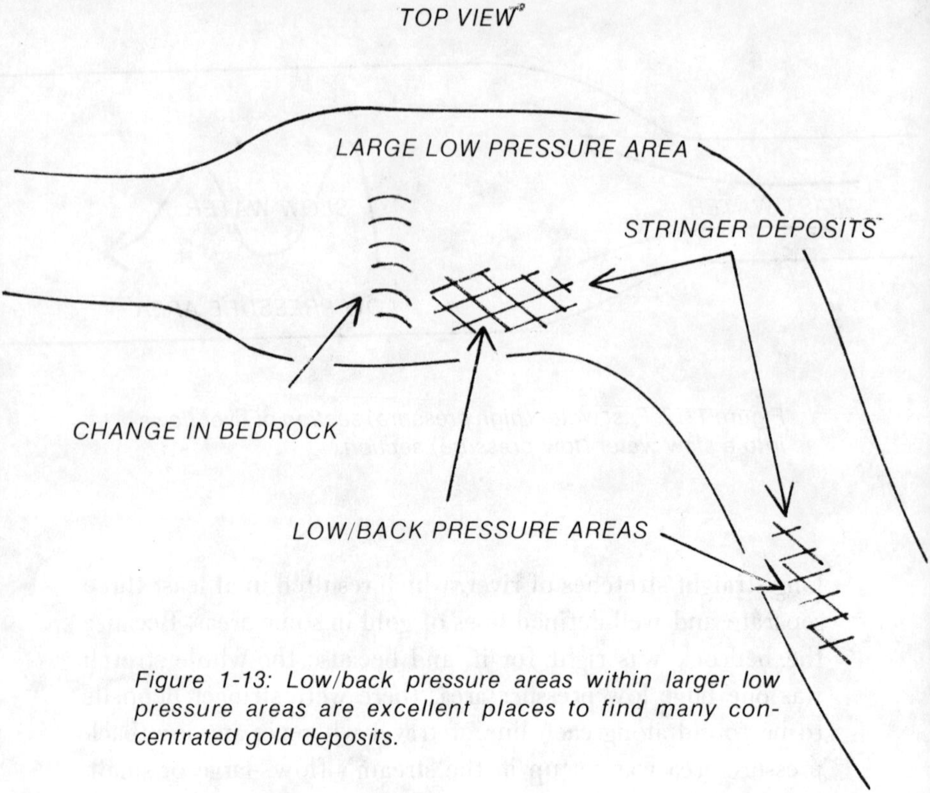

Figure 1-13: Low/back pressure areas within larger low pressure areas are excellent places to find many concentrated gold deposits.

widens for a long distance after coming out of a long narrow passage; it might travel some distance further before the overall pressure lets up enough to allow the major low/back pressures to concentrate gold instead of boiling it out. One indication to look for, which can be a big help, is where the material starts to collect in large amounts along the riverbed—like in huge river bars. There, and perhaps for some short distance upstream, is where the gold stringers ought to have formed also. Your sampling efforts will tell.

HOW THE SIZE OF A WATERWAY CAN AFFECT STRINGERS

In theory, the larger in size that a waterway is, the more stringers you are likely to find along its course, and the better formed they are likely to be. Also, the more gold they are likely to contain. There are several reasons for this: For one thing, a large river, as opposed to a small creek, carries far more force and material—and therefore gold, downstream during a major storm—as long as it is located within good gold bearing country. Because a river is so much larger than a small creek, in comparison to the size of material that is swept downstream, there will be far fewer major obstructions which seriously affect the flow of water and material. So the general line of gold's travel, and those areas of concentration, are likely to be far more orderly and well defined. In addition, the low/back pressure areas in a large body of running water are going to be of far greater size than those found in a small body of moving water. So the stringers are going to be of far greater size too.

In a smaller sized waterway, you are not likely to find too many stringer deposits. The chances are far better that you will find single deposits instead. The richer that the stream is, then the more deposits you will find in a small stream if you dredge it. Smaller sized streams, unless they are very rich, tend to be rather spotty with deposits. You will find gold for awhile; then it will just quit; then you might pick it up again, etc.

Richer rivers also carry more and better paying stringer deposits. So the larger the size of the waterway that you dredge in, the fewer deposits that you will need to locate; because they are so much larger and carry so much more gold, you will spend more time dredging them up and less time looking for them. I know of guys who have spent three and four years of dredging in the same well paying stringer deposits, and they still haven't played out yet! The one section of river mentioned earlier would have taken lifetimes to completely be dredged out by two or three dredging teams; yet in that period of time, the area will have been redeposited at least two or three times over again. All gold bearing rivers are not this rich however; but all gold bearing rivers should be found to contain an abundance of stringer deposits; and specific sections of those rivers will be found to contain more of them than other sections.

A STRINGER'S BOUNDRIES

A stringer can be of almost any width—even as wide as the river itself. In some stringers, the outside boundaries are less defined and visible than in others. Sometimes the outside edge of a stringer can end so abruptly that if a sample hole is dredged just a few feet to the outside of it, you might not see any sign of its presence. (See figure 1-14.)

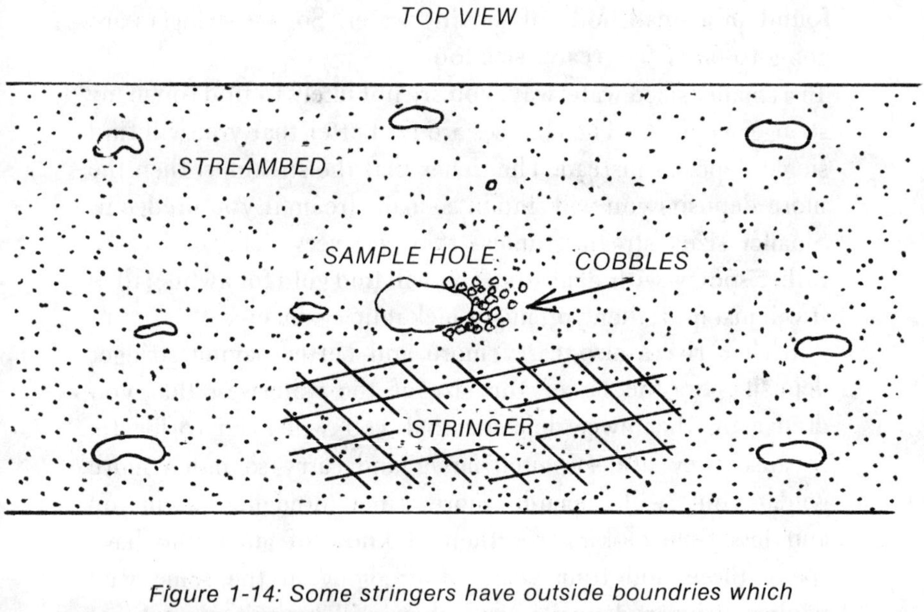

Figure 1-14: Some stringers have outside boundries which end abruptly.

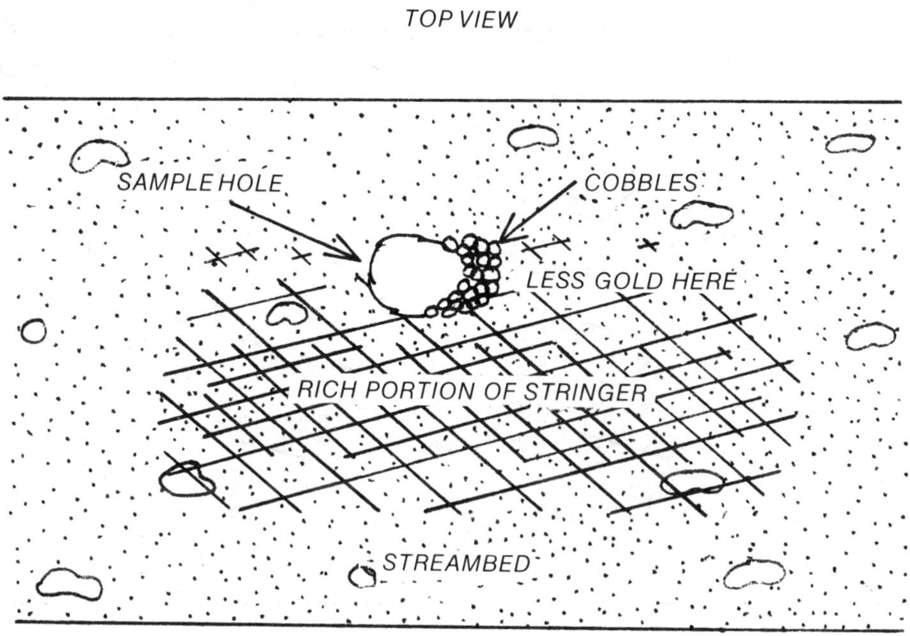

Figure 1-15: The amount of gold tends to grow less and less further out from a stringer deposit.

In most stringer deposits, however, the amounts of gold will grow gradually less as you move outward from the better paying section of the stringer. In this case, a sample hole dredged slightly to the outside of the main portion of the stringer is still likely to show some sign of its presence. (See figure 1-15.)

Sometimes a stringer will pay rather consistently all throughout. Other stringers will be somewhat spotty—paying good for awhile, then poor, then good again, etc. Sometimes the bedrock along a stringer will change from rough to smooth, and then rough again, etc., with gold only lying on the rough sections of bedrock. (See figure 1-16.)

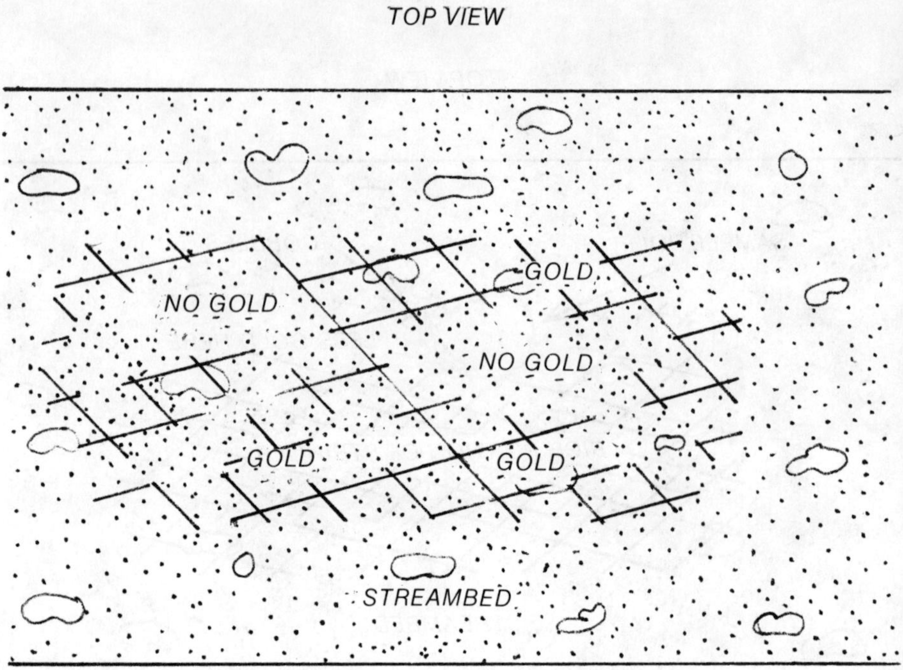

Figure 1-16: Spotty stringer deposit.

The gold in a stringer might pay only on the bedrock, or all throughout the material and on the bedrock. Or, a stringer may only pay on a hardpan, or on several hardpans, or on none of them.*

A stringer can consist of only fine sized gold, only medium sized pieces of gold (rice size), only the larger pieces of gold (nuggets), or in any combination of these. Sometimes a stringer deposit will pay in fine sized gold towards the tail end, medium sized gold in the center, and only in nuggets towards the upper end of the stringer. Or, because different sized particles of gold have different resistances to water

**For a complete description of hardpan deposits, see Chapter 4 of Volume 1 of this manual.*

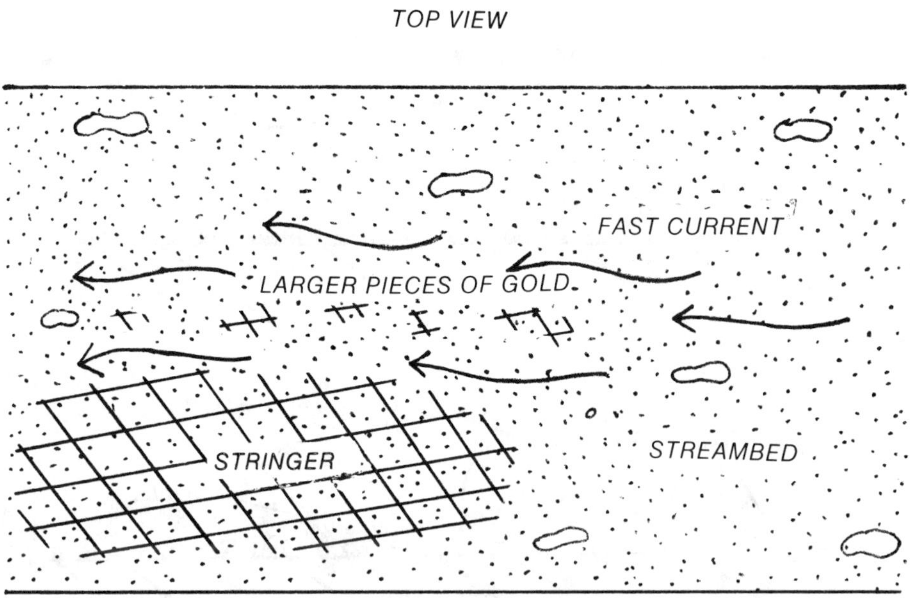

Figure 1-17: Sometimes a line of heavier gold will be found deposited just outside of the stringer deposit.

force, in some stringers you are likely to find the different sized pieces—larger ones especially—hung up on a certain line along the stringer's length. Once in awhile, you'll find some nice sized nuggets just to the outside of what appears to be the outer boundary of the better paying part of the stringer. (See figure 1-17.)

This isn't always true; sometimes all the different sized classifications of gold will be found all throughout the entire length of a stringer. Once in awhile, you will even find a good sized nugget up in the material itself, although this is more uncommon. These are usually resting on the bedrock, or on a very hardpacked hardpan.

Sometimes a stringer will pay in rather low grade amounts towards the tail end, and keep getting better and better towards its upper end. This will often happen when there is

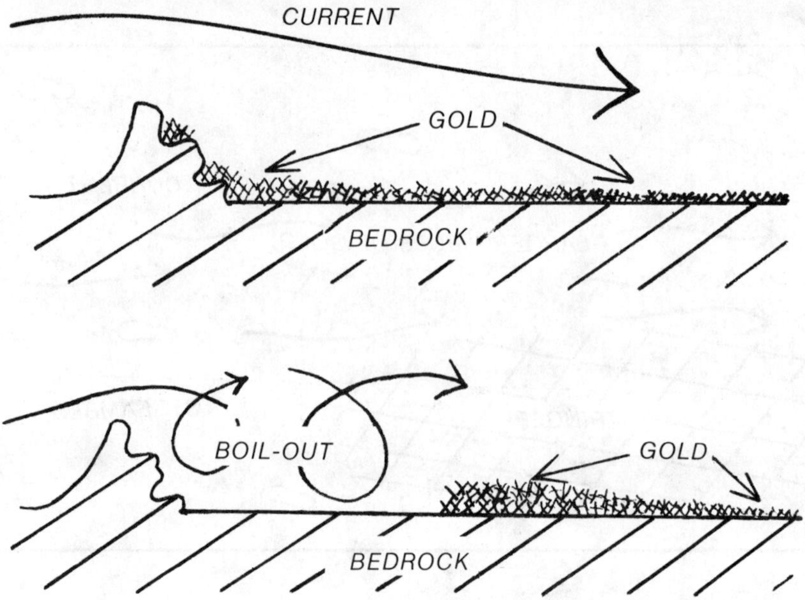

Figure 1-18: Different ways a bedrock upcropping can create a stringer deposit.

some kind of large obstruction at the upper end which is the cause of the stringer, like a large dike or upcropping of bedrock. In this case, you might find it very rich just behind and around the obstruction; or that area might be entirely boiled out. Each case can be different, as shown in figure 1-18.

Sometimes a stringer will pay rather consistently all throughout, but become spotty towards its outside edges. This can make it more difficult for a dredger to follow the outside boundaries and recover all the gold, without having to dredge up a bunch of low grade materials too. (See figure 1-19 next page.)

As mentioned earlier, stringers also contain an increase in iron rocks and black sand—especially where the stringer is

TOP VIEW

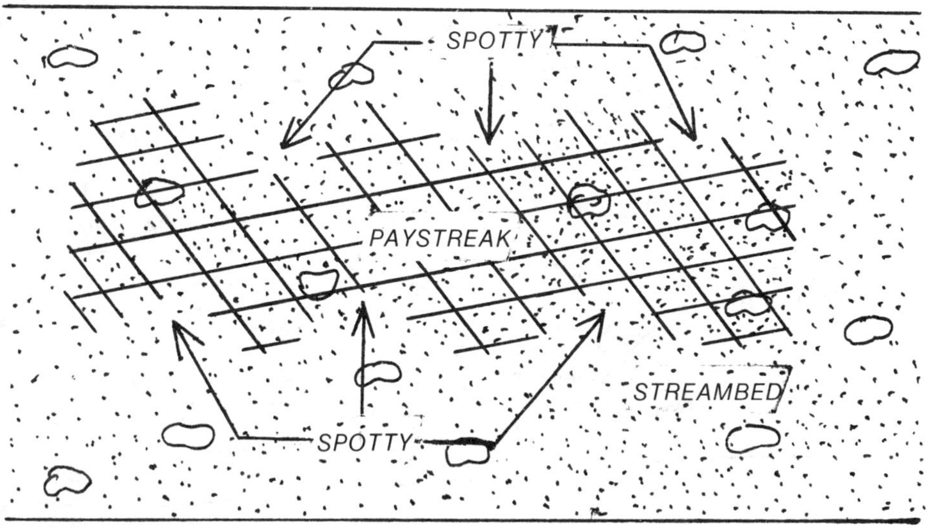

Figure 1-19: Some paystreaks are spotty towards the outside edges

paying the most in gold values, but also towards the outside boundaries, and sometimes even outside of the stringer and below it. This is especially true when a stringer has a fine gold tail end. Sometimes the increased amounts of hematite and magnatite extend out beyond the heavy concentration of fine gold, and can be detected as itself in sampling. (See figure 1-20 next page.)

Sometimes the color, hardness, and/or consistency of the streambed material will change towards the outside boundary of a stringer. This is not a true test of a stringer's outside boundaries though; because sometimes the streambed material will change inside of a stringer; sometimes there will be no change at all in the material on the outside boundary; and sometimes there will be a change in the material towards the outside boundary--but acceptable quantities of gold can

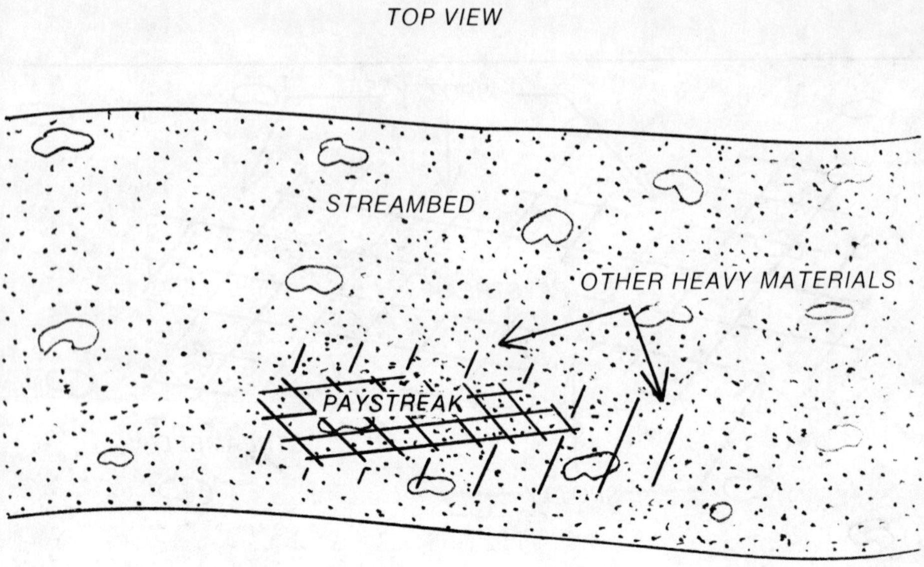

Figure 1-20: Heavy concentration of other heavy elements can extend to the outside and below a formed paystreak

still be found outside beyond the change. So the best way of locating the outside boundary of a stringer is by looking for where it stops paying in acceptable gold values.

Paying stringer deposits do not need to have hard packed material. The material can be very loose–or for that matter, there might not be any material there at all. The only real test of whether or not it's a stringer, is whether or not the gold is there!

FLAT ROCKS

I was talking to an old seasoned pro one time and he started talking about the increased number of flat rocks that you find when you get into a stringer deposit . . . And sure enough, after he mentioned it, I've come to notice this same thing myself. When a flat rock is lying down, it does not have as much surface area for water to push against as a rounder rock does. Therefore, it will take more force to move a flat rock downstream, and so the low/back pressure areas are going to show an increase in the percentage of flat rocks.

This isn't to say, however, that you are not going to see flat rocks elsewhere in the streambed—you are. Since flat rocks do not have a higher specific gravity than the average of other general streambed materials, you are going to find them all throughout the river. So seeing a flat rock by itself—or even an accumulation of them—is not a sure sign of anything, since they can be found—and even concentrated—anywhere in the streambed. But if you already have an idea of where the general line of gold's travel runs in the river, and you are punching sample holes along its length looking for the low/back pressure areas, a visible increase in the percentage of flat rocks in the material can be a good sign to look for—especially since you can't always see a visible increase in magnetite and hematite, even if it's there.

Sometimes in sampling you can find all the correct signs: an increase of iron particles, lots of lead and old rusty nails, an accumulation of flat rocks, and so forth, but no gold what-so-ever—or very little of it. In this case, you have located a low/back pressure area that is either off the main line of gold's travel, or is on the main line but no or little gold was moved through there during the major storm which formed that deposit. Also, a great deal of lead by itself in the material or along the bedrock might not signify a low pressure deposit at all, but a good fishing hole or target practice area instead. Look for the other signs of a stringer as well.

ANCIENT/VIRGIN STREAMBEDS

In **Volume I** of this manual, I thoroughly covered what a naturally formed streambed looks like and also what virgin material is, as compared to streambed material which has already been processed by earlier mining activities. All that data is correct and very valuable to a gold dredger, and it should be thoroughly studied; because a dredger is going to be at least ten steps behind if he doesn't know what natural streambed looks like. However, to give a clearer picture of the whole, more should be said about this subject, and a few clarifications need to be made.

First of all, it's important to understand that a major storm can and will redoposit new streambed–and new placer deposits–into any area which has already been gone through by earlier mining activities. Any area which caught a great deal of gold once is likely to catch a great deal of gold again if more gold is washed down into that area by a later storm. Just because a set of riffles in a sluice box is cleaned out once, doesn't mean that it won't recover gold again just as well once you start washing new material and gold over the riffles. In fact, they may recover even better after the heavy concentrates have been removed. A proven gold bearing waterway will react in the same way. So it usually makes no difference whether an area has been mined once or twice already–as long as it has not happened since the last MAJOR flood storm which occurred in the area. In fact, those areas which were once heavily mined by the earlytimers were mined because they paid well during that time period. For those areas to have paid well, they were most likely concentrated gold deposits (stringers). And those same factors which caused gold to concentrate there once are probably still present and have caused gold to concentrate there again in any of the major storms that have occurred in the area since.

So it can be greatly to today's gold dredger's advantage to know where the oldtimers hit it good. The same areas are often paying just as well or better today. Again, this is only likely to happen where major storms have heavily redeposited the streambed in those areas. But it doesn't take much sampling to find out if this has happened, as long as you fully understand what natural streambed material is supposed to look like.

An "ancient streambed," for our purposes, would be a natural streambed which has been in place since before the oldtimers arrived on the scene. In other words, those gravels and that area have never been mined before.

If you find an ancient streambed inside a low/back pressure area, along the common line of gold's travel, in a good gold bearing location, you are likely to find a good deposit of gold there too-- which will be free of old rusty nails and other man made metallic items. If you find an ancient streambed which is off of the general path, the chances of a good gold find are less likely. But all the same, it might still be worth a look, because the general path may have been different when that streambed was laid down, or there may have been an additional line of gold's travel due to some other source.

Sometimes you will also find ancient gravels lying in an area that has already been mined before, the ancient gravels having been redeposited there from somewhere else that was never mined before—perhaps out of a high bench, or maybe out of the river itself. In this case, if they are found along the general line of gold's travel, there may also be a great deal of gold along with them—especially if located in a low/back pressure area.

So there is a difference between ancient materials, virgin materials, and virgin ground. " Virgin ground," is an area that has never been mined before. "Ancient materials," are streambed materials that are either located over virgin ground, or have been swept away to be redeposited elsewhere. You can sometimes tell ancient materials by their hardpacked, colorless, decomposed texture—but this isn't always so. Any material which has been redeposited into a naturally formed streambed during a major storm could be classified as "virgin streambed, " because it has yet to be mined since having been laid down—even if it has been redeposited into a location that has been mined before.

When you find yourself dredging in a naturally formed streambed—which is not located in extremely fast water, you are dredging in materials that were deposited there by a storm or runoff of major dimensions. A storm of that size will also have redeposited a tremendous amount of gold in that waterway—some of which, if you are dredging in the right spot, could very well be inside that material or on top of the bedrock underneath. This is true even if the has been cleaned up before and you are finding signs of it, like tools and nails—especially if you are finding old tools and nails; as these things have a high specific gravity and will tend to concentrate in the exact same locations as gold does. There is an exception to this, however, and that's when the nails and tools have been washed out of the bank or some other area from not too far upstream. Just like in the case of a high bench gold deposit bleed-off, the tools may not have traveled far enough downstream to get into the main line of gold's travel, and therefore they may be on a separate line of their own.

So as far as most of today's waterways are concerned, it doesn't seem to matter if the areas have been once mined already or not. Matter of fact, those areas that once paid well, seem to be some of the better places to look today. The main thing which does matter is that the materials in those locations have been naturally laid in by a major storm.

Chapter 2

MORE ABOUT SAMPLING PROCEDURE

A hard working, persistent, sampling activity is the key to long term, continuous success in any gold mining activity of any size.

During the last several years, I have spent a lot of time running around in the general areas I've operated within—gathering information about the rivers, talking to miners, finding out where they had dredged, how they were doing, etc. And in doing so, I've had an opportunity to observe many dredging operations of different sizes and kind, some which have succeeded, many which have not. Out of all the operations I have seen fail, I would estimate that at least nine out of ten failed for the lack of having an effective sampling activity.

The philosophy behind dredging success is actually very simple; when you are into a stringer, you produce like mad; when you are not into a stringer, you sample like mad until you find one. Then you produce like mad again!

DREDGE SIZE

The size of your dredge has a lot to do with how well that it can be used in a sampling activity. It's much more difficult to have an effective sampling activity with a super dredge—unless you are mining on a very large river, where the dredge can easily be floated for long distances,with sampling being done along the way. If you are on a smaller sized river, and it will be necessary to pack the dredge in and out, in and out, etc., to sample different locations, you are much better off to have an intermediate sized dredge to sample with. It's not impossible to sample with a large dredge; and if you are dredging sample holes in deep material, you are probably far better off in the long run to pack a larger sized dredge into the area to do the job.

Packing a large dredge into a difficult area would be no problem if you knew that there was a paystreak there; but packing a large dredge in and out of different locations to see if there is a paystreak there--on the equal chance that there might not be--is a different story; it's a hassle!

The harder it is to move a dredge around, the less likely it will be moved around as much—but the success of any sampling activity greatly depends on its willingness to expend energy moving in and out, in and out, and go at it hot and heavy until a paystreak is found. This is why those who own one of the larger sized dredges should also own an intermediate sized dredge for sampling reasons—and use it. And it's also the main reason why I almost always recommend that a newcomer start off with an intermediate; because being new at the game, a beginner is going to be spending a lot of time sampling. So it is well to have equipment which can do the job—but which can also be moved around without a great deal of effort.

The five inch sized dredge is an excellent all around dredging machine. It can be moved around easily, and is also able to produce well in a wide range of conditions when a stringer is located.

To give an example of this, there's a guy I know who owns an excellent claim on the main river in the general area where I operate. The river runs with a section of fast water in the center, and slower water above and below, on his claim. At the beginning of a recent season, without having done any sampling beforehand, this miner put his super dredge onto the upper section of his claim, where he figured the gold ought to have been. He then spent most of the rest of the season dredging in that same particular spot, working his hole down to bedrock and continuing to spread it out, and out, and out—looking for the gold he thought should be there.

I saw him three times during the course of the season, myself being interested in how he was doing. Usually when you see someone dredging in the same place for days, weeks, or mouths, you figure that the guy must be doing pretty well in that spot. Not so with this guy, he was simply dredging continuously in that same spot in hopes to eventually get into some kind of nice deposit.

The worst of it was that he also had a five inch dredge floating in the water on the upper section of his claim—which was not being used at all. Each time that I saw him, I advised that he temporarily shut down the super dredge and spend a few days or a week sampling around the claim with his five incher to find the paystreak—especially below the fast water section on his claim; and once the paystreak was located, to move the super dredge there and get on with production. Each time we talked, he agreed that my plan was "probably a good idea," but he never took action on it; he just kept dredging in that same spot.

Towards the very end of the season, he finally realized that his operation had failed, and took some of my advise by sampling the upper end of his claim—only to find that there was no paystreak in that section of river. But that's something that he could and should have found out about at the beginning of the season—before even considering the idea of moving his super dredge and entire base camp into that location.

He never did bother to sample out the lower section of his claim below the fast water; the reason he said, was because it would have been too much of a hassle to move the super dredge into that section of river from where it already was located!

A short time later, he ran out of operating money and sold his super dredge to a more successful miner for about a quarter of its value. This was so he could have money to get home on. He also agreed to allow another dredger—one with more skill in sampling ways—to go onto his claim on an exclusive ten percent lease. (10% to the claim owner.)

This new miner, towards the end of last season, floated his six inch dredge upstream from a point downriver, to a position well below the fast water. His plan was to start at the lower end of the claim and work forward with a systematic set of sample holes in an effort to find the pay streak—if it was present. And BINGO!! He hit it on the first sample hole; a beautiful stringer which is paying two to three ounces/day in about 2 1/2-feet of material over bedrock. The stringer is very wide, and may very well pay in excess of 200 ounces before it is all cleaned up.

There is a lesson to be learned from this: *SAMPLING IS THE KEY TO ANY CONTINUOUSLY SUCCESSFUL DREDGING OPERATION OF ANY SIZE ANYWHERE!* and I can never say it too many times! But it's not enough to know it—you have to do it!

CHOOSING AN AREA TO SAMPLE

If you are not already onto a good section of river, or do not already have some hot leads as to where to go, one of your best bets is to locate some area which has already proven to be extremely rich in the past.

There are various methods of doing this. One way is to get hold of a *WHERE TO FIND GOLD* book, to find out the different locations in the areas of your interest which have proven to pay well already. For the reasons mentioned in Chapter 1, those areas that have proven rich in placer deposits during the past are also usually paying gold dredgers well today. Geological reports and a whole variety of maps are available to help you evaluate different general areas and specific locations in your search for a good place to start.*

Another way to find a good gold bearing location, is by contacting the *MINER'S REFERRAL,* and asking them about the general areas that you are most interested in. MINER'S REFERRAL has spent years accumulating data about the various proven locations in the United States—which is presently all going onto computer. One thing to know about this though, is that while they can be a great help—and will give you the straight dope; you will still have to evaluate the data using your own judgement; and you'll have to find your own stringers when you get to the place that you decide on. The final responsibility of putting yourself into gold, and staying into it consistently, will rest on your shoulders—always. So you should really do your best at placing yourself into as good a general area as you can. *MINER'S REFERRAL* is only one of the valuable tools which can be used to help in this. For further information, turn to page 158.

Once you do choose a general area to operate within and go there, you should spend some time running around looking the river over to evaluate which sections are most likely to pay in sizeable deposits. One of the best ways to do this is through communication with the locals and other miners in that area, to find out where gold has already been taken out by dredging activity.

It is very true that some sections of river are far richer than others. A rich section of river will be readily indicated where some paying deposits have already been taken out with dredges here and there along its length. If some paying deposits have already been taken out of a stretch of river, then it's likely that there are more undetected stringers still in place.

**For a complete outline of how to utilize geological reports, maps, and topographical maps in finding good gold bearing locations, see McCracken's GOLD MINING IN THE 1980'S. Available from KEENE INDUSTRIES.....*

It's usually no secret where the paying deposits have already been taken out. In fact, this kind of information is usually broadly known—even if it is exaggerated a few times over! Your best bet is to get your data from reliable sources—like those who recovered the gold, or if not available: by as many different sources as possible. Then, you can at least get a general idea of what happened and where. It doesn't really matter if the accounts of the amounts taken out are 100% accurate. What you are basically trying to do is locate those sections of river where you know that paying quantities of gold traveled through during the last major storm. Once you start finding out the specific areas where gold has recently been taken out, you can get a good idea of what stretches of river to concentrate your sampling activities towards.

The idea is to go for the sure thing: You pick an excellent past producing general operating area; you find out what stretches of river in that area have paid gold dredgers during the last 3 or 4 years; you go into those sections and determine where the general paths of gold are most likely to run; and then you sample the obvious and likely low/back pressure areas along that path. This procedure, if followed, will put you into gold quickly.

Actually, the only way to go for the sure thing, is when you pick up some information about a gold deposit being located in a specific place—which is still there, and you go after it if you can. This happens once in a while. Other than that, when you are sampling, you are best off to go after the most likely areas where the deposits ought to be.

There are other ways to go about finding gold: You can just put your dredge into the river anywhere and start dredging in hopes that you'll hit a deposit someday soon. I've never seen this method work very well! Or, you can go out and sample all the different streams that no-one has ever found any gold in, thinking that perhaps those were overlooked; and if you work hard at it, maybe you'll find a bonanza! It's possible—but it won't take too much sampling before you get an idea of the prospects. You could also check out those areas where streams have shown a small amount of gold, thinking that you might be more likely to find a concentration, since few miners have worked those streams. Your sampling activities will tell. There are probably many different ways to go about finding paying quantities of gold—all with a certain degree of workability. However, if you apply yourself to the method which works most often, you will be increasing your chances of continuous success by that much. In my experience, I have found, and seen, that the most consistently successful way of finding and recovering paying deposits of gold, is to go after the sure thing—or as close to it as possible.

Once I determine where the paying deposits are most likely to be. I'll make a list—with the best prospects at the top. All the items on the list will be good likely prospects. Then I'll contact the claim owner of the first prospect—if there is one, and see what kind of a deal I can work out. If an acceptable deal cannot be worked out on that section of river, I'll try to make a deal on the next best prospect on my list, and so on, until I have a place to mine legally. You will find that you can usually make a deal with most owners of claims. If they are not working the claim themselves, then they will usually be interested in allowing someone else to work the claim and give them a percentage of what is recovered. And if once in a while you can't seem to come to an acceptable agreement with a claim owner, there are plenty of other places to dredge!

GATHERING DATA

There are some gold dredgers out there who have a fixed idea that they need, or should, or have, to mine only on their own claims—even if there are no paying quantities of gold to be found there. This is too bad really, because it's far better to pay out 10%—or even 15% if you have to—of a good find, than it is to pay no percentage of no find at all—or maybe a small unacceptable one which is on your own mining claim.

There is nothing wrong at all with mining on your own claim if there are acceptable quantities of gold to be dredged-up there. If that's the case, you should be dredging right there. However, if the gold on your own claim is not all that great, or you have finished recovering the richer deposits, and you find out that there are richer paying sections of river elsewhere available to you to dredge on a percentage or lease deal, that is the way to go if you are in it for the gold and the money.

All deposits eventually play out, and that's when it's time to go find another one—whether it's on the same claim or not. And this is why you'll want to be continuously collecting new data about the whole general area—even when you are dredging up a hot stringer.

There are many ways to collect new information about the general area that you are operating within. Paying attention to where others are punching holes is one good way. If you see a dredging operation dredging a lot of sample holes up through a section of river, and the holes appear to be good ones—mostly down to bedrock, and they appear to be in the right places, and the operation continues to dredge more holes, this is one sign of experienced dredgers. So you can figure that they are doing a pretty good job of sampling that section of river, and that they did not find acceptable quantities of gold in any of those holes.

In most rivers, the water will be clear enough that you can see all or most of the sample holes which have been dredged during the present season. The following winter storms will fill in all the holes, so they won't be nearly as visible from the surface thereafter.

If you notice that an operation is continuing to dredge sample holes, it's reasonable to assume they haven't located an acceptable deposit yet.

Or, if you notice that they are producing regularly in the same spot, and you can see signs of their earlier sampling activities, you can figure that they are now into some kind of a deposit—no telling how good. But if it's not very good, you probably won't see them produce there for long.

If you notice an operation dredging in the same spot continuously, but you cannot see any signs of earlier sampling activity by that operation in the nearby areas, you really can't tell what's going on unless you go in and talk to them.

When you talk to dredgers, if you can't find out from them how well they are doing, you can at least find out how experienced they are just by asking questions and feeling them out. Then you can assess their experience (meaning their ability to find deposits) against what they are doing, and get a good idea of whether or not they are into a deposit or not.

It doesn't matter nearly as much to know how well the other operations on the river are doing, as it does to know where on the river that other dredgers are doing well. This is probably your most important kind of information: knowing exactly where the stringers are being, and have been found. When you see where others have done well, and are doing well, you can immediately see where the general line of gold's travel moves through those same sections of river; and if you have an eye for the low/back pressure areas, you can also see where other stringers are most likely to still be in place.* Any section of river that carries one stringer, is more than likely to carry more stringers in that same section.

Another excellent source of information is the local gold buyer. He will know who is finding gold, because he'll be buying it from them. Some gold buyers keep quiet about who and how much, and this is good. On the other hand, if

**Various low/back pressure areas fully outlined and illustrated in GOLD MINING IN THE 1980'S.*

you are in well with the local gold buyer, and he knows that you are honest and a good hat, you may be able to get some helpful information out of him. This is just one other good avenue of valuable information.

Another way to come about good information, which will sometimes get you onto a good claim fast, is to offer a "finders fee." If you make it known around the general area that you will give 5% of the gross find to anyone who arranges to put you into an acceptable deposit, you just might get into an acceptable deposit fast, with little or no sampling being needed at all. And after you pay it, and offer it again, you might get into another deposit just as fast, and so on. In some areas this works well.

Another good way to collect valuable information, is to talk to every active miner in the area who will give you the time of day. This is not to say you should spend all your time running around "shooting the breeze" over a beer or a cup of coffee. The only time that might be justified is when you first come into a new area and you are still trying to establish where to start sampling. Once you get started in sampling and production dredging, you are just naturally going to run into other miners in the area; and this is mainly what I'm talking about. If you can swing it, talk to them and find out what they know. If someone knows specifics, like: "Joe Blow hit a bonanza down on Big Bend last summer," and "John Doe found some great deposits down at Hurricane Flat," and if he's willing to take the time, have him show you exactly where in the river these things happened. The exact amounts taken out will probably not be wholly accurate, but that doesn't really matter; you just want to find out where all the proven sections of river are, and where the general path of gold's travel lies in those sections, and what areas have already been dredged out.

The best source of information when doing this is the miners who were directly involved with the individual gold finds. But if there is somebody there right now who is offering you information, take what you can get. Here's an example of common questions that I might ask when gathering data from the other miners in my own general area: Where has lots of gold been found on the river in past history? Where have gold dredging finds been made? Where exactly? How much gold? On bedrock? In the material? How deep is the material? What color was the paydirt? How many layers? Do they pay? Were they stringer deposits? Bleed-offs? Single deposits? Fine, medium, or coarse gold? Where else have any good gold dredging finds been made on the river? What areas on the river have been heavily dredged recently? How much gold? Percentage deals on the claims? How much? Who to talk to, and where to find them? What's the most consistently paying section of river today? Other good rivers? Places to camp? Any problems with locals, or the local government officials? Any theft or crime problem in the area?

It can be a big help to have an enlarged section of map which covers the river along with you when you are gathering data. This way you can mark in the good areas as you find out about them. This is especially helpful when you are new in an area, and are not yet familiar with the river.

It's good to try and keep tabs on as many miners, claims, and areas as possible. In this way, you will be continuously accumulating new information, and doing new re-evaluations of the river; and you will always have a multitude of "next best" sampling locations on your list of highly likely places to find paying quantities of gold.

One thing worth mentioning at this point, is that I personally NEVER withhold information about the river from other miners who ask me questions. Oh, if I know where there is a deposit that I intend to dredge out at some later date, I might not say anything about it—and I always keep quiet about other miner's deposits when I know about them. But that's about as far as my own quietness goes. I figure that anyone's success in a dredging operation depends on his or her ability to sample. The ability to sample effectively depends largely on how much data that he or she can collect about the operating area. I go about data collecting with abandon. In exchange, I give data out freely—and in doing so, I've never had any problems in finding out what I wanted to know.

"NO GOLD" REPORTS

You should not give too much credit to the information when you are being told that there is no gold left in an area. Pay close attention to where sampling and production dredging has already been done. Stringer deposits are not visible from the surface, although many of the places where they ought to be are visible to some experienced miners. Since the stringers are not visible, and since nobody has found them yet, it's easy to get the idea that they are not there. And so it's not uncommon to receive false reports from the local miners that there isn't any gold left in certain locations or sections of river. If you are smart about it, you'll pay close attention to exactly where the gold has already come out of the river, and make up your own mind whether or not certain sections of river ought to have paying deposits still in place.

Just by the nature of some stringers, someone might dredge a sample hole down a few feet off to the side of a good one, see no signs of it, and miss it entirely. It's not likely that a real Pro would come that close and miss a stringer; but some dredgers are a lot better than others, and some are downright lousy at it. Quite a few times, I've seen guys dredge right down into the center of rich stringers and miss them entirely. "Not very much gold over there," when there's a hundred ounce stringer sitting over there! An experienced dredger will start asking this guy questions about that area, the material, the gold, etc., and then go right on in there after the sure thing. I've done this a few times myself.

I've also missed my own fair share of stringers because of poor sampling procedure. I was sampling a section of river a couple years ago, for example, and floated my dredge down into a large low/back pressure area behind a large bend in the river. I knew there was gold there because someone else had been into the location before me and told me about it. Plus, all the signs pointed to a very hot stringer sitting there: In excess of 1000 ounces had been dredged out of the claim just above; the adjoining claim down below was producing just about as well; and a 20 ounce stringer had been recovered out of a much smaller low pressure area just downstream. I also knew from talking to the guy who was in there before me that the material was only two to three feet deep, that he didn't do too much work in there (he was just a recreationalist), and that his best day was seven ounces with his five inch dredge! Now that I look back on the whole thing, the signs appear so obvious that there might just as well have been a lighted blinking neon sign over the area–with an arrow pointing, and saying: "Rich Stringer Here!" But at the time the signs were not that clear to me–I just had an idea.

Well, I floated my dredge down into that spot. The water was swift, and because I was working alone–which is usually the case, it was very difficult for me to get my dredge set up into position to dredge a sample hole. I worked hard at it for about an hour: Struggling against the current, making progress, losing it, gaining it back again, and so forth; and I finally did get set up in that location. The earlier miner's hole was still there; all I really had to do was continue it for a bit just to take a look. But I was exhausted and disgusted from the exertion of setting up, didn't want to leave my dredge in that position all night, had a few hot leads further downriver, and I never did punch that hole; I floated my dredge downstream instead.

About a month later, some Pro who was operating off of the information he had received from me, moved into that location and dredged that same sample hole–the one that I didn't. He ended up pulling out better than 120 ounces of rice sized gold and nuggets inside of three weeks! My sample hole would have been right smack in the middle of the stringer; and had I gone to the extra trouble of dredging that hole–which would only have taken me about 15 minutes, that gold would have been mine. No need for you to call me stupid–I did plenty of that myself. Actually, I'm much smarter now; expensive lessons have a way of teaching a guy to do things right!

This story–which is confirmed true by me, might make you wonder how many stringers are floated right over by gold dredgers, not to mention the ones that are missed by oversight and poor sampling procedure.

So when someone says that there is no gold in an area, it just means that there is no gold in the area that the person knows about; and this is probably the best way to look at it.

There is one other thing to also keep in mind about this: There are only a small percentage of gold dredgers out there, during the time of this writing, who understand how to sample effectively–and who do so. Consequently, there are some failed gold dredgers around the various proven gold bearing areas. What better excuse is there for a failed gold dredger to use than: "There was no gold left in the river!" Some failed gold dredgers will go out of their way to convince you that there is little or no gold left in the river; for if you, being new to the area, start finding paying deposits there, what kind of a miner does it make the failed gold dredger? A failed miner of course! Yet, many of them even have an excuse for this: "You just keep getting lucky!" You are better off to not judge by the negative reports of failed miners.

You can get far more accurate data from the continuously successful miners, but even the negative reports from them are suspect; maybe they don't want others moving into their general area of successful operation! People are just not continuously successful when there is "no gold," and this is something worthwhile for you to know. Just find out what's already been done, and found, and where, and use your own judgment. That's the best bet.

EXAGGERATED DATA

On the other hand, you also have to watch out for exaggerated information. There have been many times when I've gone to town and heard reports of my own recent gold finds which had been blown up to three, four, or even as much as ten times larger than the finds actually were! This happens; it's all part of the game, so you have to expect it. As mentioned earlier, the exact amounts do not really matter nearly as much as knowing that it was good, and pinpointing the exact location of the find.

Sometimes the trouble with exaggerated data, is when you hear of a "Tremendous," "Incredible," "Amazing" find from another general area some distance away, it tends to make you want to "rush" on over there to start dredging. Before deciding to pick up and abandon the general area you are operating within for a new general area, it's a good idea to try and substantiate the reports to the best of your ability. Who knows–the reports that you are receiving about that distant area might be small compared to the reports that they are receiving about your finds in your own general area of operation!

There is nothing wrong with leaving one general area for another if it is a better one. However, it usually takes being in an area for a while before you gather enough understanding of it to where you can evaluate accurately where the deposits are likely to be, and start putting yourself into them rather continuously. This is a lot to give up because of an unconfirmed report or two!

Also though, there is the chance that after having been in an area for a while, your greater understanding of it may show that there are a few deposits to be found, and you'll need to find a new operating area. So don't shut out those rave reports and success stories either; they might be just what you need!

SAMPLING PROCEDURE

How many sample holes that it will take to get into a good paying stringer deposit depends entirely on how good the general operating area is, how well you place your sample holes, and your thoroughness in following up positive signs while sampling.

It might take a few more holes to find the first stringer in a certain section of river, than it will to find the additional stringers in that immediate area. Once you find one deposit, after it's cleaned up, your sampling is likely to turn up additional stringers in the likely spots along the same line of gold's travel in the same immediate area.

When dredging a sample hole in a likely spot, it's usually a good idea to dredge one or two more off to each side, to make sure you do not miss the stringer if it's present. This includes taking a sample of the bank if it's right there, and also dredging a hole or two out into or towards the gut (center of river)—especially out into the gut—even if the water is a bit swift out there, as long as you can accomplish it safely.

You need to keep your sluice box relatively clean of gold while dredging sample holes. Sometimes you might pick up a few pieces of gold out of a sample hole that you might not see while dredging. If you already have a bunch of gold in your sluice box, you might not notice the difference. Sometimes just a few additional pieces of gold in your box may be the only sign that you are sitting right next to a paying deposit—even if it's of bonanza proportions. So it's very important to be able to tell when you have ANY amount of additional gold in your box.

Sometimes you will accumulate some gold in your box from earlier sampling in areas that had some gold, but not in paying amounts. If the accumulation becomes enough that it will interfere with your ability to see additional gold in the box—even in small amounts, you are better off to dump the concentrates off into a tub, and dredge them up again later when you are ready to do your next clean-up.

A few pieces of gold seen on the bedrock, in the material, or ending up in your sluice box after dredging a sample hole, is one of the best signs to look for of a nearby stringer. This is especially true if the pieces of gold are larger than fines.

If you are seeing an increase of gold in a sample hole than what would normally be expected in the general material outside of a stringer deposit, your sample hole should be moved to both sides and forward, with you looking very closely to see if you can find more of it, and see in which direction it seems to get better. A few pieces of gold seen while dredging a sample hole are ALWAYS worth checking out further.

If you dredge a sample hole and do not happen to notice any gold, but find some new gold in your box after the dive, it's always worthwhile to go back down and open up the sample hole further, while looking closer, until you find out where the gold is coming from and how much is there. With me, many times the smallest clues have led to the richest pay streaks!

If in dredging further, it doesn't seem to get any better, dredge a few more holes in the immediate area to see what else you can find. It may be that you are near a good paying stringer; or it may be that you are already inside of a low-grade stringer; or it may be that you are along the general line of travel—and not into a low/back pressure area yet.

Sometimes with further dredging in the same sample hole, you will see more gold on the bedrock, or in the material; but you still can't tell if the area is rich enough to dredge on a production scale. In this case, it can be worthwhile to spend a dive or two, or the rest of the day, production dredging in that spot, and then clean up to see how good it is going to pay. This is best done after additional sampling in that immediate area has already shown that you are not going to find anything much better around there. Sometimes this happens when dealing with widely dispersed fine gold which is spread throughout the material. It can be so fine and spreadout, that you can't see much of it; because it's so fine. This happens, and sometimes it can pay quite well to continue on there.

Some rivers are so heavily saturated with fine gold that you have to clean out your sluice box quite regularly while sampling, to observe any increase after each sample hole.

Sometimes, once you have finished cleaning up a stringer, after your first sample hole, you may find some gold in your box that you are reasonably sure did not come out of the hole. After working in a good paying stringer, some gold can become trapped in the classifier inside your header box. It doesn't usually take too long for this gold to work its way free and then arrive in your riffles—this usually happens on the first sample hole or two. However, it can be misleading to you while you're sampling if you are not aware of it.

When choosing places to dredge your sample holes, pay attention to where the large scale earlier mining operations took place. Many of those old-time operations washed a lot of gold into the river—particularly the hydraulic mining operations. Watch for where their tailraces entered the river, and do some sampling right there if you can. When these do pay, they usually pay very well. Sometimes the gold from these tailings has been washed downriver in a major storm to form an additional common line of travel, separate from the main one which runs down the river. The same holds true for high bench bleed-offs.

If you watch for these places and dredge a few sample

holes downstream of them where you judge that the line of gold might run, you can find yourself dredging up stringers and single deposits along the "wrong side of the river"—sometimes even on the outside bends! Usually though, when you follow a line of gold to the outside of a bend, you'll find single deposits of the larger sized pieces of gold where the changes in bedrock and other obstructions allow for them. Outside bends, for the most part, are lousy place to locate stringers, because there is such an increase of force there in a major storm. This isn't always the case though; if it is a mild bend, and conditions set up a low/back pressure area there, and there is a source of gold emerging into the river not too far upstream—like the remains of an old hydraulic mining operation, you just might find a well developed stringer on the outside. I have seen one such instance of this myself.

So it's not just the deposits on the main common path that you are necessarily sampling for. If one of these other conditions presents itself, it's worth an extra sample hole or two to check it out. "Nothing ventured, nothing gained," is a truism in a sampling activity.

READING MATERIAL

As mentioned earlier, sometimes the gold will end abruptly at the outside of a stringer. More often though, the amount of gold keeps getting less and less further out from the outside edge of the richer portion; and so there are usually some signs that a stringer might be present nearby, if you dredge your sample hole to the outside and not too far away from the stringer. If you see these signs, they can usually be followed right into the stringer deposit.

So keep your eyes open when you are dredging sample holes. Not only are you looking for gold, but you are also watching for those places where you are most likely to find an accumulation of gold if it's present. A change of layers is one example of a good place to find gold if it's going to be there. The same holds true for bedrock; you'll want to be looking the the places on the bedrock where gold would definitely catch up if a stringer was present in that location. You uncover these particular areas slowly and look closely for any accumulations. These are the most likely places to find gold if it's going to be there.

You are also looking for accumulations of rusty objects and lead in these same places–which are a good sign that you are dredging your sample holes in the right locations–or near to them. Increases in magnetite can be a good sign too–although you might not notice this until you finish the dive and look into your sluice box.

In some rivers, some colored materials pay much more often than other colors–like a red color. This is not always true; but if it is, it just gives you more to go on–and look for.

If you start seeing an accumulation of the heavier minerals, rusty objects, or lead on the bedrock, or on top of a layer, or in the material, you should start spreading your hole and watch closely to see where there seems to be more of it, where in the material it is located, and see if there is any gold present too–this is the best sign. Once you start seeing gold, you look closer to see exactly where it seems to be coming from out of the material. Then you start looking to see in which direction it appears to get more plentiful. And you move your hole in that direction, always uncovering the area where it seems to be paying the most slowly enough that you can see more if it is there. And pretty soon, if it's there, you'll be into the stringer deposit.

DEPTH OF MATERIAL IN SAMPLING

I have run across more than a few dredgers who believed that the material had to be very deep in order to pay well.

This is simply not the case. Most of my own recoverys have been made in three feet of streambed material or less. I have seen, or know of, literally thousands of ounces of gold that were recovered out of three feet or less of streambed material.

I have also seen a lot of gold recovered down at the 10 to 25 foot levels (of material).

All rivers seem to have their deep sections and their shallow sections of river. Every river that I have been on has had its deep deposits and its shallow deposits. In fact, I recovered a stringer of black sand and fine gold last summer in Alaska which ran right down across a section of exposed bedrock. There was no material over it at all. That stringer was very easy dredging, it paid quite well, and had I not seen it myself, I might never have believed it!

The deeper that a stringer is located under streambed material, the better it will have to pay for you to recover the same amount of gold, with the same production equipment, on a daily basis.

It seems best to direct your sampling activities towards those areas that you can effectively dredge with your production equipment. If your effective production capabilities are in six feet of material or less, why dredge sample holes greater than six or eight feet deep? One good reason would be if the sample hole was being put down in a prime location and all the signs of an excellent deposit were very positive. Once in awhile you'll find yourself in this predicament, and you will have to use your own judgment when deciding how deep you are going to dredge a sample hole before pulling out.

I can usually dredge three or four good 3 1/2-foot deep sample holes in the same amount of time it would take me to dredge an effective nine foot deep sample hole in hardpacked material. So what is better, an increased amount of effective sampling at shallower depths? Or, to sample fewer areas at greater depth?

Certainly if you can't work the deeper sreas effectively, then you should concentrate on covering more of the shallower likely spots—but what if your sample hole is going deep, but all the signs are very positive? This is something you'll have to decide on yourself, because each situation is different. You have to be careful not to get pulled too far into a single sample hole though. To finish a sample hole just because you started it, is not good enough reason to go well beyond your effective production capabilities.

I'm quite certain that there must be some gold bearing rivers which run deep for the most part, and so most of the deposits will run deep too. You'll find this out fast enough once you start sampling and collecting data about the area. If this is the case, and you don't have the equipment to effectively handle the depth, you are better off to choose a different general operating area where the rivers have more shallow areas to dredge. There are many many of these to choose from.

Deep areas are easiest sampled with larger equipment. When figuring out the amount of time and energy it would take to dredge a 25 foot deep sample hole with a five inch dredge, you probably could not go to too much trouble to haul an eight inch dredge into the location to sample it with.

So when looking over and gathering data about a section of river, you should also be very concerned with material depth.

It's good to have a system of how deep to take a sample hole before pulling out because it's going too deep. Different dredgers have their own methods of determining this, and you'll have to decide on your own. Just to give you something to go on, I'll tell you what I do: I figure that I can dredge effectively down to about six feet of streambed material with my six inch dredge. I can dredge deeper than this, but it has to pay very well to make it worth my while—mainly because I find so many stringers at less than half that depth. Since I am comfortable dredging in six feet or less, I'll dredge a standard, safe, good sized, orderly hole down to about the 4 1/2 to five foot level of material. If at that point, there is still no sign of gold—or of bedrock, I'll make a dive straight down for bedrock at the bottom of my hole. By this I mean dredging a small hole—about two or three feet in diameter—straight down, simply to see if I can find bedrock. This will usually give me another two or three feet of depth, unless I run into a large boulder that prevents my further passage downward. But if I do find a large boulder, it tells me that the material goes considerably deeper; and this is what I'm trying to find out anyway. If I do not run into a boulder, and if I still haven't touched bedrock, or run into good gold at the 6 1/2 to seven foot level, I'll pull out of the hole and go start a new one elsewhere.

There may be some experienced dredgers out there who do not entirely agree with this method, and that's fine. There are different ways to go about this; and even the best method in one location, may not work very well in a different location. This is something you'll need to use your own judgement on in your own particular areas.

No doubt, I've missed a few deep deposits by pulling out of sample holes in this way; but I've also found a lot more shallow deposits for having covered so much more area by effective sampling!

If you do not have some method of determining how deep you will take your own sample holes, it's possible to get over involved with a single sample hole to the point where it can become a major production operation just to reach bedrock. And of course, the whole idea behind sampling is to avoid getting into a major production operation until AFTER you find a good paying deposit.

To show a good example of this, a couple of seasons ago

there was a guy who came to ask my advice about his operation. He had a VERY LARGE home built eight inch dredge downriver on his own claim about four miles from where my base camp was located. His operation consisted of four other guys and himself. They were hard workers–keeping the dredge operating during most of the daylight hours. They were not recovering any gold to speak of though, and that's why they were interested in my advise on what to do.

In looking over the situation, I found out that they had not yet hit bedrock, but they had been at it full time for over a month! Their hole was 18 feet deep; but they were in loose packed material, which kept sliding in on them–especially at night. All the cobbles and boulders had to be carried or winched out of their hole. This was a very slow process.

My advise was that they move the dredge to a new location–which they were dead set against doing; because they had already spent so much time, energy, and money in that single hole, that they couldn't bear to let it all go to waste!

They also had a three inch dredge on the claim which they had brought along for their wives and kids to play with. My second advice was that they put at least one of their guys full time on the three inch dredge, sampling around the claim to see if better and shallower ground could be located to place the super dredge. This, they thought, was a joke; all of them being certain that the three inch dredge was simply a toy to be played with.

So they continued on in that same hole for the rest of the season, and they hit a packed boulder layer at the 22 foot level. The boulders were too large to be moved by hand, and too tightly packed to get a boulder harness around for winching. All they could do was spread out the hole in hopes to find an opening down to bedrock. They never made it; the first winter storm filled their entire hole back in with the same loose packed material.

So there went an entire season, two dredges, five full time operators, not one completed sample hole, and no gold. Those guys didn't come back to dredge last summer!

How deep do you take a sample hole before pulling out because the area is too deep? It takes a bit of judgment; you'll have to decide.

SAMPLING DIFFICULTIES AND HOW TO OVERCOME THEM

The hardest part of sampling is not the hard work involved with dredging lots of new holes down to bedrock—although this is tough work in its own right. And it's not in having to haul your dredging equipment into and out of the various new locations—which is hard work too. Nor is it in having to work out new deals with the owners of the claims that you want to test out. It's not even a combination of all these things that makes sampling tough to do. The hardest part of sampling is in not knowing where the paystreaks are located, or how long you are going to have to go before you locate one. The main problem in sampling is in overcoming all the unknowns!

If you knew where the paystreaks were surely located, negotiating the deals, moving the equipment around, and dredging the holes would all be easy to do—just so you could start dredging up the gold! In a sampling activity, you have to do lots of hard work in an effort to reach bedrock, in specific locations, to find gold—which you are not certain is even there in the first place; and it's the mental barriers that need to be overcome, just so you can continue to sample effectively until a paystreak is found.

Starting a hole is far more difficult than continuing one which has already been established down to bedrock. A lot of the time, you don't even know how deep you'll have to dredge a hole to reach bedrock—which is just another unknown that needs to be contended with.

Sampling basically consists of starting new holes, and taking them down to bedrock—or as far as your judgment tells you to go, and continuing to start new holes, and holes, and holes, in this manner, until you find a paystreak. None of this would be too difficult if you knew where the stringers were. It's not knowing where the stringers are—and all the hard physical work piled on top—which makes sampling tough for some gold dredgers.

It gets especially tough if you just dredged ten or 15 holes, having hauled your equipment into and out of four or five different locations, and you still haven't located any pay dirt yet.

I've seen many beginning dredgers punch one single sample hole and then give up because they didn't find gold. I've seen others give up before they even finished the first hole! Sampling takes a lot more persistence than this. You have to be willing to stick with it. Like I said, it's a mind trip not to mention the hard physical work involved too.

I myself am not immune to this problem. Sometimes if the going gets real tough, I have my own doubts and uncertainties to overcome—just like anyone else. I felt like calling it quits a few times—but that was before I found enough paying stringers to gain my own certainty that they are out there in abundance, and that they can easily be found just by mechanically persisting with standard procedure.

Tough as it may be for some, it should be realized that to succeed well at gold dredging on a continuous basis, you must have an effective and energetic sampling activity.

In a production activity, you will recover as much gold as you are putting time and energy into it; and you tend to find yourself really pouring on the steam—at least I do. And in doing so, you get paid handsomely; you are seeing an immediate exchange for all your hard work; you feel great about it; and you don't mind putting in the long hard hours—because you are dredging up treasure and riches!

Seeing all that gold being sucked up into the nozzle tends to keep you going beyond all normal endurance—especially when there are lots of hand-pickers being uncovered as you are moving along. This can keep you pretty excited.

If you don't approach sampling with the proper state of mind, in putting out a lot of time and energy, and in not receiving a direct and immediate exchange for it, you can lose a lot of moral fast. First of all, a production activity can be somewhat accurately gauged by how much gold is being recovered. A sampling activity cannot be guaged in this same way; and if done so, you will generally find yourself slowing down. When little or no quantities of gold are being recovered, incentive and momentum can be lost.

Probably the best way to measure how well that a sampling activity is doing, is by looking at how much ground is being tested by thorough and effective sampling.

If the stringers are present, they will be located quickly by a thorough and energetic sampling activity. If they are not present, that will be found out quickly; and you can move to a new area which shows better promise and continue on there.

So in sampling, if you put your frame of mind onto producing like mad to cover as much prime area as possible with good sample holes in all the right places, instead of putting your attention on all the gold you're not getting, you will get much more done, find more stringers, and manage to keep the morale up all along.

Sampling is pure detective work, and to do it right, you have to be willing to do it as thoroughly as possible—based on your judgment of what needs to be done. You have to be energetic in following up those clues.

I know of two guys who have dredged a great many sample holes, with at least 85% of them having never been completed. It's not that most of their holes were particularly deep or difficult to do. They just didn't have the endurance to complete their sample holes, and would decide: "There's no gold here," after starting many of their holes, before many of them were even taken very far. This was not a physical endurance problem; I never once saw them lying down on the streambank from physical exhaustion! This was some kind of a psychological attitude that they had worked themselves into about not being able to find any gold—or something like that.

Many of their holes—at least the ones that I saw—were in the right places; and had they spent the additional 15 or 30 minutes, or an hour or two, to complete each job, they certainly would have struck pay dirt. As it is though, they have yet to find a single stringer.

Not too long ago I ran into one of them in town, and he was all upset because some other miner had finished dredging one of their abandoned sample holes and had pulled out better than 50 ounces in a few weeks of dredging. That's about 100 times more gold than those guys have recovered to date! "God, if I'd only finished that sample hole!" People who give up so easily carry their own failures around with them—and not just in gold dredging!

Now I don't want to get too aloft about this; we all have our moments of uncertainty and weakness, and we all goof up once in a while. I just want you to know that lots of persistent sampling, thoroughly done, will get you into acceptable paystreaks. After you have had some success at it, you will gain your own certainty. Meanwhile, you'll have to just go out there and do it.

There are many guys out there who know how to do it, and who are all ready to talk about it, or to dabble at it lightly, but who just can't seem to get out there and do it hard. There is something about a sampling activity that makes it tough for some guys to apply themselves fully to the task. It's easy to find yourself "too tired" or "too disgusted" to do the job thoroughly. One good example of this is the earlier account of the stringer I missed and someone else recovered; because I didn't dredge a sample hole where I should have, and he did.

One way to get around this is to look the area over ahead of time and decide on all the spots where sample holes should be dredged, and mechanically stick to that plan no matter what—unless through sampling, you discover that additional holes are needed too. If you find yourself becoming too burnt-out to do a thorough job of it, it might be best to take a day or two off, and then go back and hit it hard right where you left off. Probably the best thing to do though, if you can, is to just push on through with it; because the sooner you dredge the holes, the sooner you'll find the pay dirt, and the sooner things will be great again.

It has been said that intelligence can be defined as the ability to estimate the correct amount of effort required to overcome the barriers in completing a task. It takes so much effort to carry a filled gas container from the truck down to your dredge. It takes more effort to carry your dredge from the truck down to the river. If you use too little effort on a project, you probably won't complete the task. If you use too much effort, the project can be goofed up—or at least some of the effort will be wasted.

In sampling, because so much physical activity is required, and because the main barriers to overcome are your own unknowns, there is no way that you can use too much effort to accomplish the task. So give it all you've got; hit it hard man; enthusiasm, devotion, velocity, quickness, thoroughness, vigorous, industrius, perseverance, movement, hustling, human dynamo, eagerness—this is the best way! The stringers are there and you'll find them if you try hard and stick with it. And remember, as you continue to gain experience, it continues to get a lot easier.

FALSE CERTAINTIES

After hearing me talk about some of the difficulties in sampling, you might get the idea that sampling is a long and drawn out process. It can be, if you don't do a good job of it, or if you have put yourself into a lousy operating area. (One good index of the operating area is: are any other operations finding paying quantities of gold?) For the most part though, it usually does not take very long to find the next stringer. It's not knowing how long, which causes most of the problems—and these problems are pretty well going to have to be dealt with until a stringer is found—even if it only takes one hole!

Once a stringer is found, you tend to get all excited and forget about all of the uncertainties that you felt during sampling. In fact, once you find a stringer, you can look back and see just how easy it was to find it. After the paystreak is found, the reasons why it is there become obvious. I've had so many guys come up to me after I'd located a good deposit and tell me how obvious it was that the stringer was there, that it's not funny—so obvious, in fact, that to have missed it would have been stupid! But if the presence of those stringers was so obvious, how come someone else didn't recover them before me? The reason is because it's a lot harder to see where they are with certainty before you find them.

Once you get into a stringer, all the problems disappear; you are making excellent money; the world looks bright; you can look back and see just how easy it really was for you to find the deposit; you can feel just great about your sampling abilities, and how easy it's going to be for you to find the next ones, etc. But don't let this fool you too much. When the stringer plays out and you haven't gathered any new sure-thing hot leads, and it's time to go out and FIND a new stringer, all the unknowns kick back in again; and you tend to start worrying about "if" and "how long," etc.

If you are going to continue to play the dredging game, it's necessary for you to prepare yourself to handle these kinds of extremes. One day you are not making any gold, and the next day you are making thousands. Then after a while, the stringer will play out and you are not making anything again while you are looking for a new stringer to dredge up. It can be a tough thing to handle when a long rich paying stringer plays out; you can get the idea that you will never find another one as rich or better. I have personally found that there is always a better stringer which can be found through more sampling–and there are always plenty to be found that are good enough. Although I have seen a few dredgers quit after their first stringer played out, because they were not able to cope with this severe kind of change.

Regardless of all this, while the dredger is still dredging up the paystreak, you will never find a guy who is more confident about his ability to find more of them. I have especially seen this in guys who have never sampled before, but happened to luck into a deposit right off the bat. I am often amazed at how far off the mark these guys can be about how they are going to find the next deposit! Reality has it's own way of bringing false certainties like this down to sea level; but until this happens, you have to watch out; because a guy with false certainties–who is making good gold–can be very convincing. This is the main reason why I say to beware of taking advise too seriously from a dredger who has only found one deposit.

The true test of a dredger's prospecting abilities, is whether or not he can find his own deposits one right after the next. And of course, if he can, then he is certainly worth listening to.

Chapter 3

GOLD DREDGING PROCEDURE

DEFINING ACCEPTABLE GROUND

It's a very good idea for you to define for yourself what amount of gold is minimally acceptable to you on a daily basis. It might be a half ounce per day; it might be more; or it might be less—depending on the size of your production equipment and your operating expenses. The main point here is that you should set some minimum amount, which if you are making on a daily basis, you will stick with it; and if your daily averages start dropping below it, you'll start sampling around for better ground.

There are several good reasons for doing this; one being if you don't know exactly what is and what isn't acceptable, you can waste a lot of time "going on hoping" in some very low grade material, instead of "going on looking," like you should. Another good reason is you are going to need to determine how far to the side and how far to the rear to dredge a stringer into the lower grade material when it's present. (We'll get to this later in this Chapter.) Perhaps the best reason that you need to define a minimum acceptable level of daily gold in your own mind, is so you can take in the better, and much better, and tremendously better pay dirt as a bonus when you find it—which will be quite regular if you are good at sampling. If you do not do this and take on the good finds as a bonus, you can find yourself comparing the good finds with the acceptable finds, and pretty soon the acceptable finds will not be acceptable any more!

I know a guy that was new to gold dredging, who was into a very large paystreak, which was paying him a little more than half ounce per day for every day that he went out and dredged. This, to him, was great. That's more than $200 day CASH profit he was making over top of expenses, and gold prices are going up to make that even better. That's not bad wages, and he was really happy about it until one day when he uncovered a bedrock upcropping and pulled six ounces in that single day. The following day, however, he was back into the half ounce amounts again, only that was no longer good enough; he quit shortly thereafter. He left the area, never came back; and I haven't heard of him since!

If you stick with gold dredging long enough to get good at sampling, there will always be greater highs; but if the greater highs are the only thing that is now acceptable pay dirt, you'll soon go broke!

If you define for yourself what is minimally acceptable as pay dirt, as you continue to get better at sampling you will find that there is a surplus of acceptable ground available to you. So you will be able to upgrade your minimum acceptable levels a degree or two as time goes along.

If you are willing to dredge up every bit of acceptable ground that you can get your suction nozzle into—and are willing to accept the bonuses as they are uncovered, you'll be making more gold more of the time—and you'll find more bonuses too.

DREDGING UP PAYSTREAKS

There are basically three different phases to dredging up a stringer—all of equal importance: finding the stringer (sampling), locating the tail end and outside left and right boundaries of the stringer, and dredging up the entire stringer.

One important thing to know about dredging up a stringer, is that you really only have one chance to recover the whole thing without having to go to a lot of additional effort—and that chance is on the first try.

The first thing to do when an acceptable paystreak is located, is drop back and find its tail end. To do otherwise and throw a lot of cobbles and dump your dredge tailings on top of a portion of good paying ground is very poor planning; it just means that you will have to eventually go back and move them all over again, if you are going to recover that gold during the present season. Finding the tail end of a stringer can be twice as hard to do if there is twice as much material sitting on top of the paystreak because of your tailings—more so actually, because dredge tailings are loose; and when you are trying to dredge a hole down through them, they just keep on sliding down into your hole.

It's much easier to drop back right at the beginning, locate the tail end, and start dredging from there; but, as much sense as this makes, you really have to discipline yourself to do it. When you find a new stringer and start seeing all that new gold, you can very easily become mesmerized by it, and want to continue on to see more of it, to see how much is there, and to pick up as much of it as you can quickly. The more gold you see, the more you want to see how much more is there, and so on it goes. Meanwhile, your tailings and cobbles are piling up on good--or even better—gold bearing ground downstream from you.

I expect that most dredgers will have to dump a whole bunch of tailings on top of a good paying stringer or two before they gain a full understanding of this for themselves. That's the way it was for me at least—like I said, expensive lessons teach you to do things right!

Any time spent in finding the exact tail end, and the outside left/right boundaries of a stringer right at the beginning, will be time saved three or four times over—or more, before you are finished with that stringer. Plus, you will recover a lot more gold. Even if it's a very large stringer and it takes a full day or more to discover the tail end, it's worth your while to do it.

Quite often, as you drop back towards the tail end, you will notice that it is not paying as well as it was where you first found the stringer. This is very likely to happen since you will want to drop back as far as you possibly can—as long as it still pays above your basic minimum acceptable daily production requirement. This can be hard to do if your minimum is a half ounce per day, and you uncovered an area up ahead of you on the stringer that is paying ten or twenty times that much—nuggets and all, which you are moving farther and farther away from as you drop back. "Discipline! You have to overcome your lust for gold, my son!" And if you don't believe in gold fever, or feel that your personal discipline is strong as nails and cannot be broken, wait until you get into the position of needing to drop back and find the low-grade tail end of a stringer when you have located what appears to be riches up ahead of you!

One good way to look at it, is that the dredging is certainly going to improve as you move forward on the stringer! This tends to spur you on quite a lot; you end up getting more done faster—just so you can get back up to where you were seeing better amounts of gold. So, you end up making more in the long run.

There are two different ways to find the tail end of a paystreak: back-tracking, and dropping back while dredging holes. If the streambed material is relatively shallow over top of where the gold is located, and the water is not moving too fast in that section of river, one good way is to track the gold back to the tail end of the stringer. This is done by placing your dredge slightly downstream and off to one side of where you found the gold, and by dredging backwards—meaning WITH the current, instead of against it—and follow the gold along until it plays out. (See figure 3-1.)

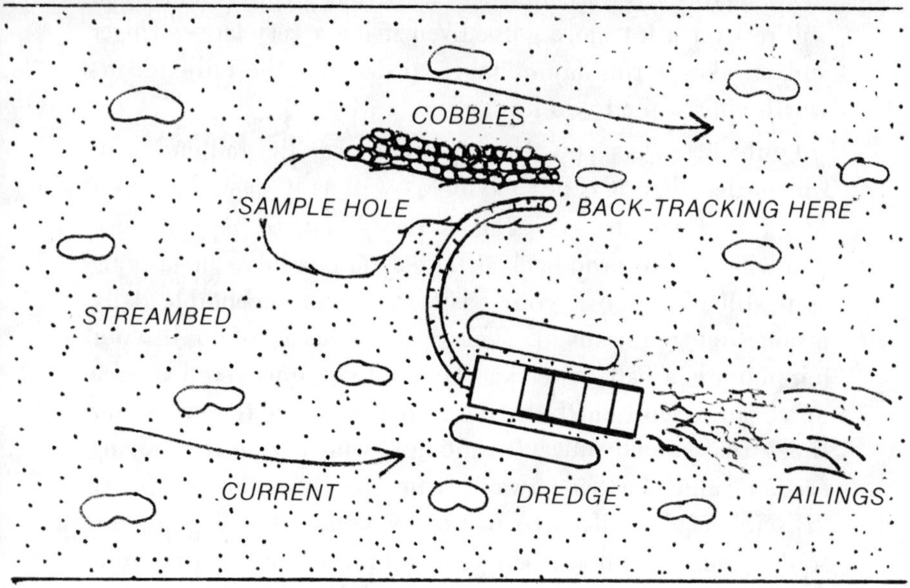

Figure 3-1: Back-tracking with the current in a stringer deposit to find its tail end.

When "back-tracking" along a stringer in this way, your main purpose is not to recover all the gold that you uncover; it's to move backwards as quickly as possible to find the exact tail end of the stringer. You'll recover ALL of the gold once you turn around and start moving forward again. Meanwhile, the more material that is disturbed while moving back along the stringer, the more misplaced cobbles and tailings that you will need to deal with once you do turn around. So ideally, back-tracking is accomplished as quickly as possible, doing no more than is necessary to follow the line of gold back to where it quits paying well enough to dredge.

For the most part, if the gold is in the material or on the bedrock in any acceptable amounts at all, you will be seeing plenty of it while you dredge. In most stringers, the gold will be found concentrated along a common area--like on the bedrock, or between two layers. At the head of a stringer, if the gold is being found in a specific layer, you are most likely to find it in the same layer towards the tail end of the stringer.

Sometimes the material changes and you can't find the same layers as you drop further back on the stringer. In this case, it might be necessary to do a little more looking around back there until you find out if the gold is there, and if so, where it's located now. This will give you something new to look for as you drop back further.

If the material is too deep over top of a stringer to back-track it easily, or the water in that section of river is too fast for you to turn around and be dredging along in the direction of the current, then instead of back-tracking, you will have to use the other method of finding the tail end of a stringer—which is to keep on dropping back, dredging new sample holes as you go along. This is more time consuming than in back-tracking; because you cannot follow the gold by sight to it's tail end. Instead, you have to guess where the tail end might be; drop back to that point, and dredge a hole; find that same area in the material that was paying well up forward on the stringer; look the area over closely to see if it's still paying—and how good; and if it's still good, to discipline yourself to pull out of the hole, drop on back farther, and try it again, etc. This can be really tough to do when you have to keep pulling out of good gold in front of you, to start new holes behind you; but there really is no other way that I know of, if you want to dredge up ALL the gold.

Once in a while, you can find a stringer in sampling, drop back to find the tail end, and not find any gold back there. Then, as you move forward, you find out that what you found in sampling is all there was—a single deposit! This happens, and can be kind of disappointing—but it's all part of the game.

When you are attempting to find the tail end of a stringer, each time you drop back to dredge a new hole, it is well to set up your dredge off to the side, where you estimate that the dredge tailings will be off the stringer and out of your way when you come back up through.

Sometimes a stringer will be spotty towards the tail end—or it may be a spotty stringer altogether. This is something to watch for while back-tracking or punching holes to find the tail end. If you are back-tracking a good showing of gold and it suddenly ends, drop back further and punch a few quick holes to see if you can pick it up again.

If you find a low-grade section of streambed material along the stringer, and the water is moving slow, that might be a good place to start dumping your tailings so you can start dredging forward towards the higher-grade material—even if the stringer does pick up again further downstream. In slow water, dredge tailings will just pile up. So if they are dropped on a low-grade section of riverbed, they will give you no further trouble downstream. (See figure 3-2.)

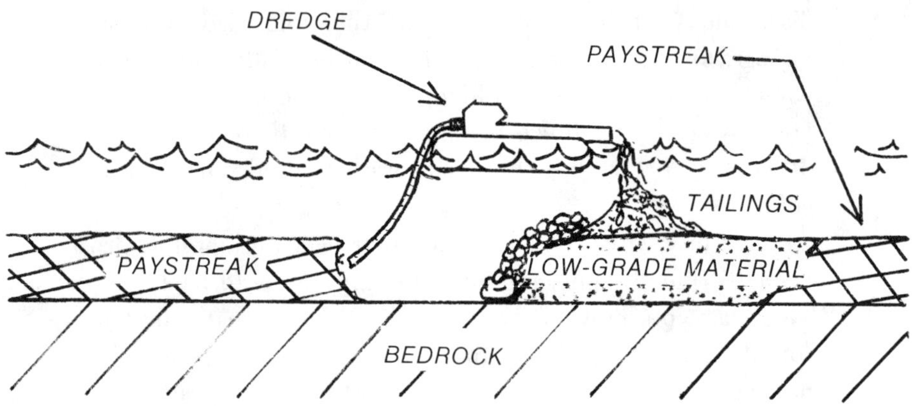

Figure 3-2: Dropping tailings on low-grade material in slow moving water

Fast water will sweep a good portion of your tailings downriver, and will have them sitting in abundance on top of pay dirt if it's located just downstream along the same line of where you are dredging. Since stringers usually run along the same line, and sometimes quite close to each other, in fast water it's usually worthwhile to dredge a few sample holes downstream, to make sure that the acceptable pay dirt does not pick up again where your tailings are going to end up. In the long run, it is best to know at the beginning what you are going to be dropping your tailings on top of. All paystreaks eventually play out, at which time it's necessary to go find the next one. A lot of the time, the next stringer will be found just upstream--or just downstream. If it is just downstream, and you have just dumped tons and tons of tailings all over it, you are going to be very disappointed!

Once you have found the tail end of the stringer, and have a good hole started, your next important task is to find the outside left and right boundaries.

Ideally, the entire stringer from left to right should be taken forward all at once. Sometimes a stringer can be so wide that this cannot be effectively done. If this is the case, the stringer will need to be dredged up in two or more separate sweeps, starting from one outside boundary and working towards the other. When doing this, be careful to keep your tailings and cobbles well away from the unworked section of pay dirt off to one side, so they won't be in your way on the next pass. (See figure 3-3 next page.)

One of the best ways to locate a stringer's outside boundary, is to get into gold and follow it outward to the side–just like in back-tracking. If you know where the gold is coming out of the material, and watch closely as you follow it out toward the side, you will see where it stops paying, and there you have found your boundary.

In that a stringer will sometimes pay very well along one part of the stringer, and pay less and less as you move further off to the sides, it's important to keep your minimum acceptable daily production requirement strongly in mind

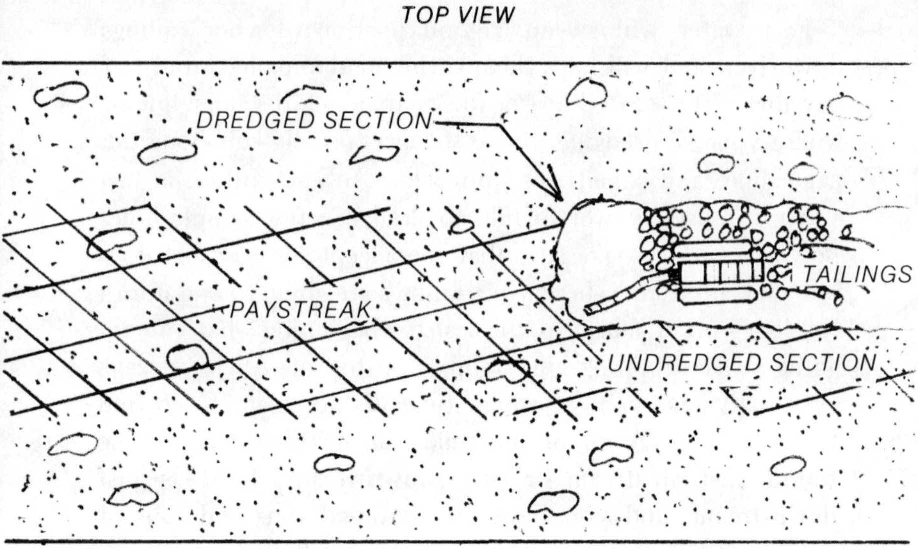

Figure 3-3: Dredging up a paystreak in two separate sweeps — dropping tailings away from undredged section

while taking apart the stringer. Otherwise you can get so excited about uncovering the richest portion, that you can end up leaving behind large portions of lesser grade material–which is still in acceptable range; and it will eventually lie under your cobbles and dredge tailings as you move forward. (See figure 3-4 next page.)

It doesn't hurt too much to dredge up the material a foot or two to the outside of where you feel the stringer's outside boundaries lie. In this way, you can be reasonably sure that you are not going to leave any good pay dirt behind as you move forward.

You can also go overboard by dredging too much valueless material to the outside of a stringer. This can cut severely into the amount of gold you bring home daily, and is wasteful also. The idea is to make sure that you dredge up all the acceptable pay dirt, and not too much of the valueless

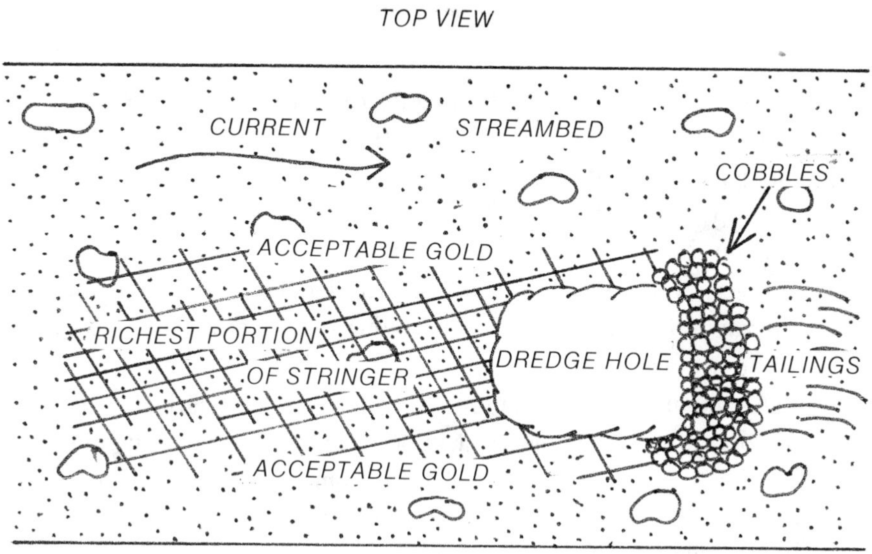

Figure 3-4: Dredging just the richest portion of a paystreak — leaving acceptable portions behind.

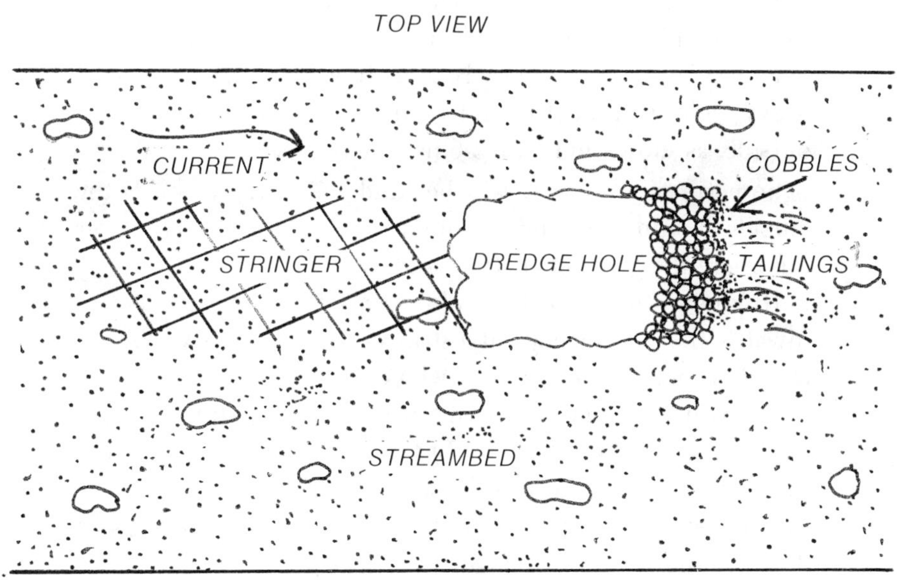

Figure 3-5: Dredging just to the outside of acceptable gold in a stringer — as you should.

material, if you can help it. You do this by keeping your eyes wide open as you dredge towards the outside edges. Look for the gold there. It will not take you long to SEE exactly where the boundary is. Then you dredge just slightly to the outside edge of that point as you move forward on the stringer, watching closely as you dredge along that line for any signs that it might be improving over there. (See figure 3-5 on last page.)

As mentioned in Chapter 1, sometimes a specific line of heavier gold (larger pieces) will be found deposited along one outside edge of a stringer. This is something to keep your eyes open for also, as you dredge along the stringer boundaries. I have noticed that when this special line of heavy gold does exist, it usually lies along that boundary which is closer to the river's gut and faster water. A lot of the time, the water can be moving pretty fast out there; and there may be some swept bedrock with material trapped in the bedrock irregularities. Those pockets of material are worth looking into, even if they definitely appear to be outside your stringer; because sometimes that special line of heavy gold will be found to run a short distance to the outside, as shown in figure 3-6 next page.

Sometimes a stringer will be kind of spotty towards the outside boundaries, and so you'll have to watch even closer for where the gold gets better—and follow it.

If you notice that the gold seems to be getting better towards the outside edge of a stringer, investigate—ALWAYS. On occasion, you can find a whole new part of the stringer which you would not know about unless you follow up this kind of sign. (See figure 3-7.) Once in a while you will find this out and need to drop back again to find the tail end of the new section of a stringer, to prevent dumping tailings on it as you move forward. (Figure 3-7 next page.)

THOROUGHNESS: That is the proper state of mind when taking apart a stringer, always being interested in exactly what the stringer is doing. You don't know until you look.

TOP VIEW

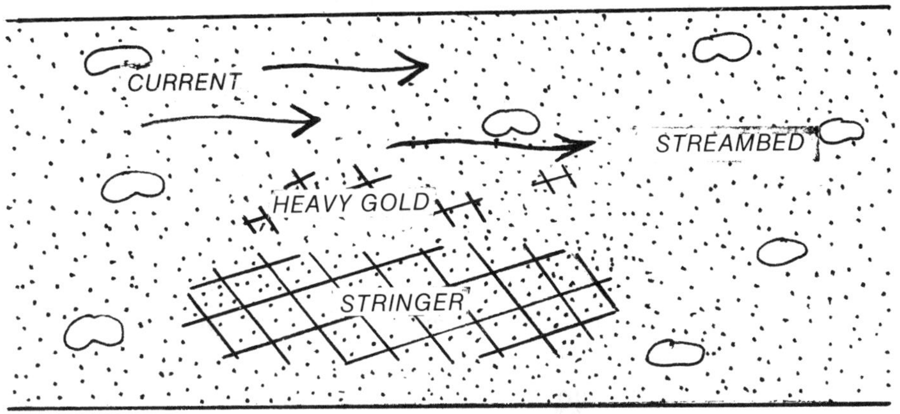

Figure 3-6: Sometimes a line of heavy gold can be found just to the outside of a stringer in the fast water.

TOP VIEW

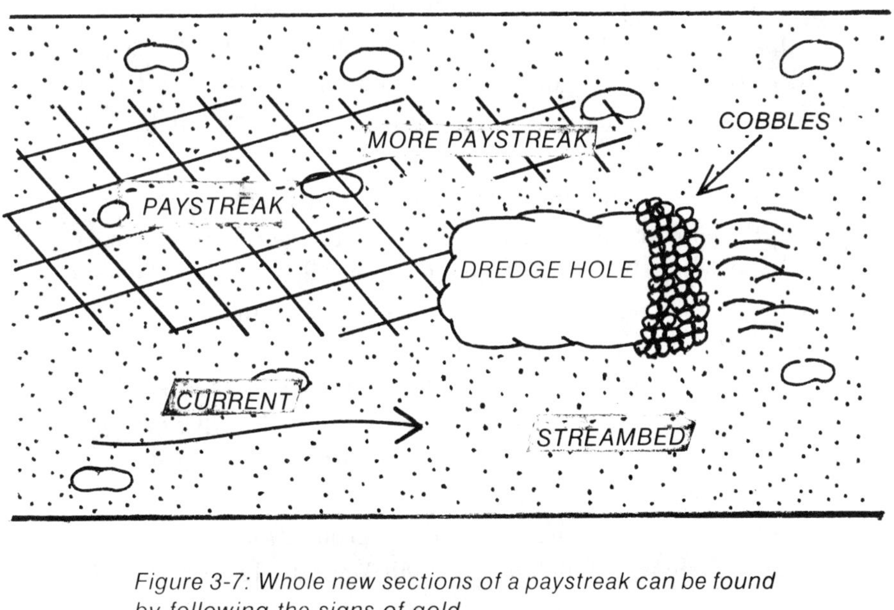

Figure 3-7: Whole new sections of a paystreak can be found by following the signs of gold.

You can go overboard on looking too, however; you can spend so much time looking, that it slows you down to the point where you are not dredging production amounts of streambed material. When dredging in a stringer, the more pay dirt that you move across your sluice box, the more gold you are making. So the only time you really want to be slowing down to look closely, is when you are uncertain what is happening with the stringer and it's necessary to figure out what's going on—like when establishing exactly where an outside boundary is located. There is no need to slow down to look when you are in the meat of a stringer and know it. You'll have plenty of time to admire all that gold later—and there will be more of it too, if you don't let it's presence in the material slow you down too much while you are dredging.

Actually, you can take apart a stringer in such a way that you can see most of the gold without it slowing you down. Once you know where the gold is coming from out of the material, you can uncover that area or layer at production speed and take notice of the gold as it is uncovered, before you dredge it up. This is something that you will quickly pick up on once you start production dredging in a stringer deposit—and it's a good way to go about it; because it gives you a continuous look at what the stringer is doing, where it's paying the best, etc.

Sometimes the streambed material will change in color, depth, consistency, or hardness towards the outside bounary of a stringer. When this happens, do not assume that the gold ends at that point too. Sometimes it does, but sometimes it doesn't; you'll have to look and see.

Once in awhile, you can run across shattered or decomposed bedrock which can be saturated with fine gold, medium sized gold, or even nuggets down inside the cracks—and way down inside of the bedrock. In this case, if doing so meets the minimum daily production requirements that you have set down, you will want to break open the bedrock and get into it as deeply as it pays to do so. If the gold is there in acceptable quantities, you will be seeing it as you break open the bedrock. Sometimes this can really slow down your forward moving progress on the stringer, and you will have to force yourself to do a thorough job of it because of the better pay that rests on top of the bedrock and in the material further forward on the stringer. You have to do it right though, the only other alternative being to waste it by leaving it behind—which you will inevitably regret having done once the stringer plays out.

I have seen bedrock in some places pay very well in fine gold as far as three feet below the normal bedrock surface. It was shattered bedrock, all full of cracks, that was underneath a fine gold paystreak which rested on top of the bedrock. It was obvious that the bedrock was paying well down inside; because I could see the gold as I broke apart the cracks.

Some bedrock doesn't pay worth a darn—even when all cracked up or decomposed. In this case you will not see the gold—and it will not end up in your sluice box.

SPOTTY STRINGERS

Some stringers pay rather consistently all throughout, and these are very easy to dredge. Other stringers are spotty all throughout, and these can be more difficult to dredge. A spotty stringer tends to pay good, then lousy, and then good again, and so on—not just as you move forward, but from side to side too. And so it can be more difficult to keep track of the exact outside boundaries in a spotty stringer. (See figure 3-8 next page.)

The only way to find and recover all the gold out of a spotty stringer is to dredge the whole thing up, taking the good with the bad. This can be done if the daily averages meet your minimum requirements. If they do not, then you'll have to sample around for richer portions and take them as they are found. This is relatively easy to do in

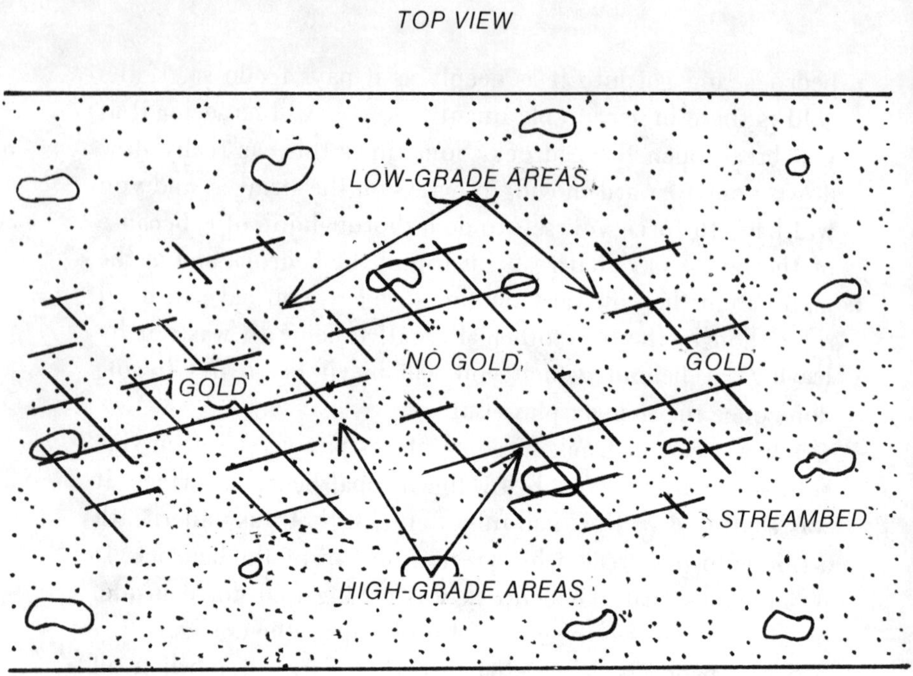

Figure 3-8: Spotty stringers are more difficult to follow.

shallow material, but becomes increasingly more difficult as you need to go through deeper material with your sampling.

One of the main reasons why a spotty stringer will go hot and cold is because of changes in the bedrock. If you are dredging along on a bedrock stringer (gold lies on the bedrock), and suddenly the bedrock goes from rough to smooth, often the gold will disappear altogether until you find the right kind of bedrock again in that stringer.

When it appears that you have dredged to the upper end of a stringer because it has stopped paying, one good way to see if it will start paying again up forward, is to place yourself in line with where the richest portion of the stringer was running; and do some fast forward-tracking—just like in back-tracking—only in a forward direction to see if you can find if and where it picks up again. (See figure 3-9)

TOP VIEW

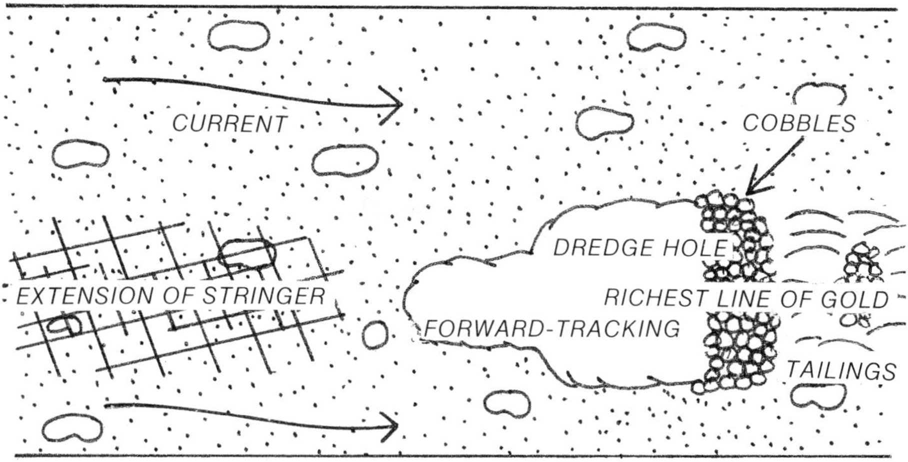

Figure 3-9: Forward-tracking to find an extension of the stringer.

TOP VIEW

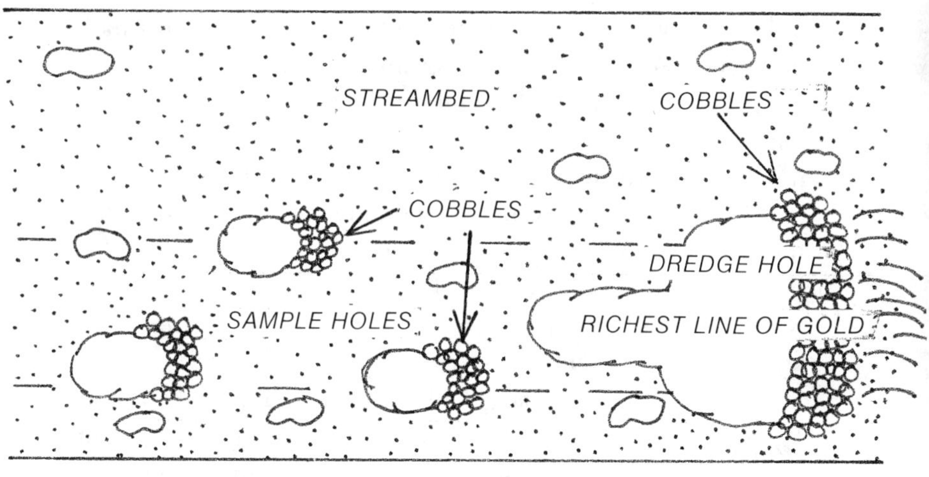

Figure 3-10: Dredging sample holes forward in line with the richer portion of a played-out stringer to find an extension or a new stringer.

If this doesn't get you back into more gold, then start punching regular sample holes forward in line with where the richest portion of the stringer ran. This will usually put you into an extension of the same stringer—or into a new one if it is present. (See figure 3-10 on last page.)

If the gold is paying on a certain type of bedrock, and then quits because of a change to smooth bedrock, as you move forward with sample holes, you are looking for the correct kind of bedrock again. Once you find the right kind of bedrock, you look to see if you can pick up the line of gold. There is no use in spending a lot of time looking for the gold if the bedrock is wrong.

If you cannot find an extension of the same stringer just upstream, start sampling further and further along the same line of gold's travel to see if you can pick up another low/back pressure area. This should be done further downstream too. Where one stringer is found, there will usually be more nearby.

FINE GOLD STRINGERS

In a fine gold stringer, sometimes the gold can be so widely dispersed throughout the streambed material, that you cannot see much of it while dredging—even though there might be paying quantities present. In this kind of a stringer, it's extra hard to tell where the outside boundaries are exactly located. Pretty much, the only way to accurately test the pay dirt in this type of stringer, is to dredge up a fair sized portion of it and then look into the sluicebox to see what kind of increase in gold has been accumulated. Sometimes you even have to clean-up to tell for sure.

When I am into a stringer of this kind, I'll spend the first 15 or 20 minutes of each production day dredging a portion of streambed along that boundary which I am most uncertain about, and then I'll look into my sluicebox. Since I clean up the sluice every night, any new visible accumulation of gold will have come out of that first section of streambed that I dredged up. In this way, I can guage pretty well where the outside boundaries of the stringer are located.

When using a triple sluice type of recovery system in a fine gold stringer, keep a close eye on the center box. I have found that the side boxes recover fine gold very well; but the center box—having much more water velocity moving through it, naturally will not hold the fines as well. Most of the fine gold ends up in the side boxes anyway, but some does get into the center box too. When I am into a rich fine gold stringer, and notice the center box accumulating a lot of gold values, I will normally dump it off into a tub (just the center box), continue on dredging, and then dredge those concentrates back up again at the end of the day, before cleaning up. When you dredge up a set of sluicebox concentrates like this, be sure to suck them up slowly—with your dredge operating at normal production speed. Otherwise, if you try dredging them up with your dredge running at a slower speed, there might not be enough suction power to pull the heavy minerals and gold up through your suction hose. If not, the concentrates will load up in the hose, and when you finally do realize this and turn up your dredge, the concentrates will all be sucked up into your sluice box at once. This will overload the riffles, and you will lose gold as a result.

WASTING GOLD

One thing to know about fine gold and lower grade material, is that sometimes you can think it's not that valuable when you compare it to the richer portions of a paystreak—and the nuggets when they are present. The thing to keep in mind about this is that gold is an extremely valuable commodity, and so it does not take very much of it to amount to some value. For example, at today's rate of exchange, one grain of gold is valued at a little more than a dollar. This means that a fourth of a grain would be worth the equivalent of a quarter in today's early 1983 buying power. It doesn't take much of a flake to amount to a fourth of a grain in weight. Yet, because flakes of that size are so small, you can find yourself easily overlooking them—even when they are abundantly present in the streambed material. If all those flakes could be magically turned into nickles, dimes, quarters, and dollar bills, etc., I'll bet a lot fewer of them would be left behind by dredgers who are more eager to get at the richest portions of the paystreaks! In actual fact though, those small and tiny flakes of gold have far more intrinsic value than today's coins do, and this is going to be more so as time goes on.

"IT'S ALREADY BEEN DREDGED!"

Just about every good gold bearing general area has specific locations or stretches of river that were once very rich, but are reported to have already been dredged out. It is seldom true that these locations and stretches of river have been thoroughly dredged—even though the locals may believe that they have been.

A lot of the time, in the "rush" to get at the richest sections of pay dirt, cobbles and tailings are dropped on top of the richest sections and other high grade portions of pay dirt. This is even more likely to happen when more than just a few miners were in there mining the location at once.

It takes years and years to thoroughly dredge out even a very short stretch or rich paying river—unless the material is very shallow and there is little of it. If the entire river was not dredged from bank to bank, then some portions were left behind. Since those sections of streambed were not dredged up, who knows for sure what amounts of gold are down inside of them? Those portions might be rich—or even richer—than those portions that were already mined.

Up until this point, I personally have only run across a small percentage of dredgers who know how to, and who are willing to thoroughly clean-up a stringer, and who know how to find additional stringers that lie along the same common line of travel. It's possible that the dredgers who reportedly cleaned out those exremely rich sections of river were not fully in the know about these things, and so left some excellent ground behind. In fact, it's likely that they did!

Also, when a lot of guys are mining in an area over a period of years, they tend to lose track of who dredged what and where. When tailings are dropped on a section of river, it's not likely that anyone is going to look underneath those tailings—especially when there is great gold being uncovered further upstream. The winter storms will blow all those tailings out of there, but by that time everybody figures the area has already been cleaned up . . .

A lot of this sort of thing happens when a number of dredgers are working in a hot-spot stretch of river, and a great deal of gold can be left behind. It's the "rush" to get at the best pay dirt before someone else does.

Me? When I see an inexperienced guy find nice pay dirt, I'll always take note of where his production hole was started. Nine times out of ten, he will not drop back to find the tail end; and nine times out of ten, he will not go back and finish what he missed; because his tailings are in the way! Inexperienced dredgers usually leave just as much gold—or more—behind them as what they dredge up. I wasted my own fair share during my early days too, and I have cleaned up after a few guys since.

So when you hear that a very hot stetch of river has already been dredged out, the guy who is telling you probably believes it—even if it's not true.

When going in to check out such areas, there are ways to quickly find out which portions of streambed have already been dredged, and which portions have not been dredged.

Winter storms usually will have already filled in earlier dredge holes; but most of the time if you can look at the river from just the right angle, you can see where the undredged portions of natural streambed stand out from the less firm dredged places. When they are sitting right next to each other, the surface of loose packed, filled in material looks different than the surface of undisturbed portions of natural streambed. A lot of the time, the difference will remain obvious even years after the holes were dredged and filled back in. (See figure 3-11.) Unless it is located in a shallow and extremely fast water section of river, dredged material will not reform into a natural streambed formation until the next major storm comes along—which will redeposit the whole area anyway, so it won't matter what was already dredged. Meanwhile, there is likely to be a visible difference from the surface which the dredger can look for when searching for the undredged portions of streambed in a heavily worked stretch of river.

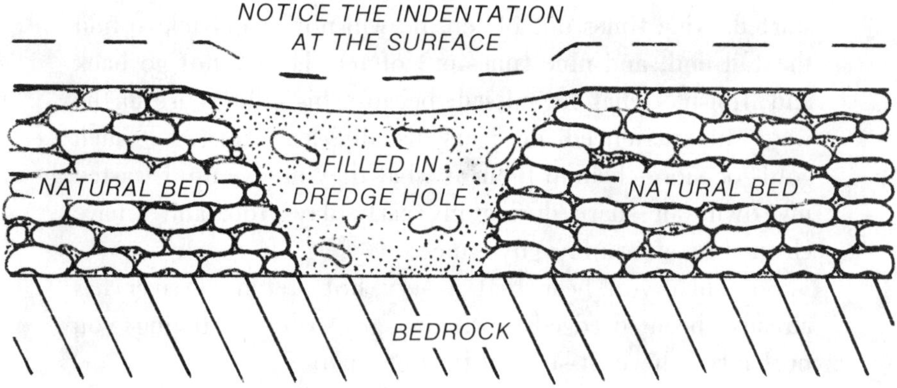

Figure 3-11: The surface of dredged material looks different when sitting next to natural sections of streambed.

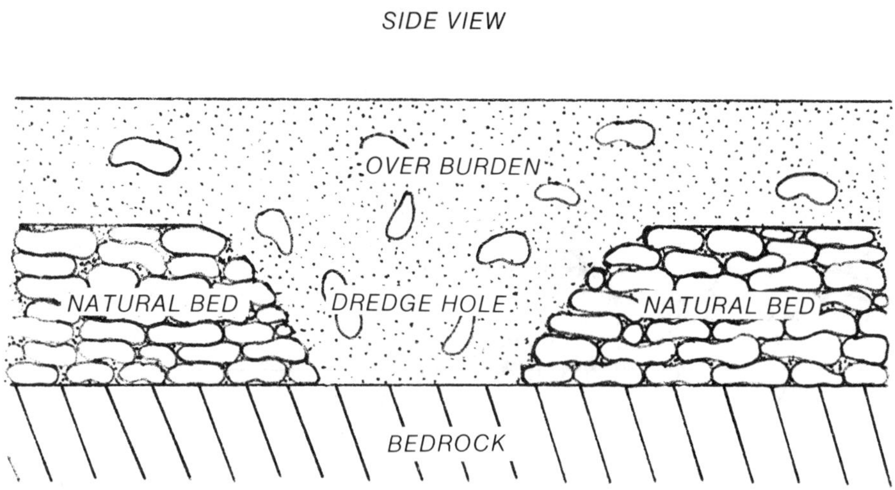

Figure 3-12: Overburden can hide signs of undredged sections of streambed.

If the undredged portions are not easily recognizable at a glance, you can put on your face mask, swim around, and pull cobbles out of the streambed from different locations here and there. The natural streambed should be more hard packed than the dredged material; and when cobbles are pulled loose from the bed, a deep colored silt is likely to be released. A filled in dredge hole probably will not do this—and if it does, the chances are very slim of its slit being the same—or even similar—to the natural bed's silt. The color and hardness of the natural bed should be quite different than the dredged portions of streambed.

In some places, there can also be a layer of loose packed overburden resting on top of the dredged and undredged sections of streambed, as shown in figure 3-12. This can make your job of finding the undredged portions more difficult; because the overburden, even if shallow, is likely to obscure the above signs from surface detection. Your best bet in this case, is to dredge a systematic set of sample holes. These should not have to go any deeper than where the overburden should leave off to natural streambed when it's present, as shown in figure 3-13.

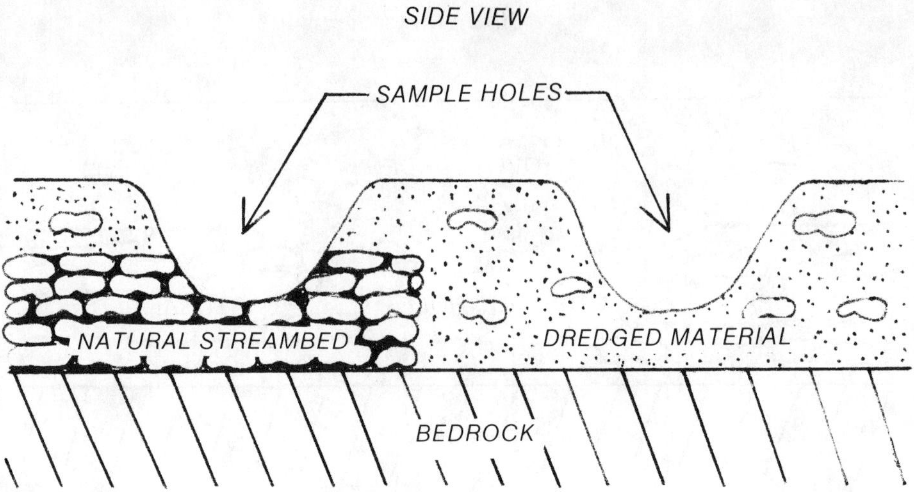

Figure 3-13: When looking for undredged streambed, sample holes only need to be dredged as deep as where the bed would be located if it's there.

When looking for undredged portions of natural streambed in this way, if you are uncertain what the natural portions look like in that area, it's well to find some in a place which has certainly not been dredged–like off to the side somewhere. Then when you start dredging holes, you will know exactly what you are looking for.

One of the best areas to find unworked pay dirt in the "already dredged out" stretches of river, by the way, is in the fast water.

DREDGE SIZES–AND SUCTION HOSE

Under most conditions, in experienced hands, a larger sized dredge (hose diameter) will move far more material than a smaller sized dredge will be able to move. For example: I can move at least twice as much material through a five inch dredge, as I can with a four inch dredge under most conditions; I can move at least twice as much material through a six inch dredge, as I can with a five inch dredge; and I can move at least twice as much material through an eight inch dredge, as I can through a six inch dredge, in many situations.

Larger diameter suction hoses are generally also thicker walled, heavier, and less flexible, than smaller diameter suction hoses. Therefore, while a five inch dredge can move twice the volume of material, it's also about twice as hard to move the standard five inch hose around, compared to the four inch hose. This same thing holds true as you go on up to larger sized standard hoses. The larger they are, the heavier and tougher they are to handle.

Some guys are strong enough that they can operate an eight inch suction hose EFFECTIVELY by themselves all day long, five days a week; most guys are not! If you are not strong enough to do this, but if that is the size of dredge that you own, you will be physically burnt-out and miserable a lot of the time; and it's likely that you would end up putting more material across a sluice box in the long run with a smaller sized dredge—one that you can effectively run for long hours without it running you into the ground.

The reason why I keep mentioning "standard suction hose," is because I'm referring to the types of hoses that usually come with store bought dredges, and which are usually available from most mining equipment shops. Thinner walled hoses are available if you look around. These will be lighter, more flexible, and therefore: easier to dredge with—especially in the larger sized diameters. For example, I've seen eight inch suction hose around which has the same wall thickness as the standard five inch suction hose, and which provides about the same flexibility as a standard six inch hose. This is quite an improvement, and makes operating an eight inch dredge much easier to handle. Some of the dredge shops are starting to offer thinner walled hoses, and I expect that we will see more of this as time goes along. If you are considering the idea of buying a larger sized dredge, you might consider ordering it with thinner walled hose—it usually costs less too!

Thinner walled suction hose will probably not last as long as thicker walled hoses; but when dealing with the larger sized dredges, so much maneuverability is gained, and therefore so much more gold can be made in shorter periods, that the lessened life span of the hose can fail into insignificance.

There are many different kinds of thin walled hoses available—you will have to look around. One thing to keep in mind though, is if you use a clear one you will have a much easier time of finding the plug-ups while dredging.

Also, some hoses can be too thin and will collapse or wear out in no time at all—you'll have to use your own judgment.

I saw one guy up in Alaska last summer who was using very thin walled air vent type hose on his ten inch dredge. The hose was wire wound and so thin that it was lighter, and had greater flexibility, than the standard five inch suction hose. It was almost like a slinky! He told me that at first he had trouble with the hose collapsing on him. Then he wrapped a long length of rope around the outside of the hose along the inside grooves between the wires, which he said completely solved his hose collapsing problem. I never did get a chance to try it out; but he was out there producing most every day, and he was never shut down because of suction hose problems. I'm sure there are other ways to go about this also, the main point here being that the thinner walled larger diameter suction hoses will allow you to dredge more volume and work longer hours.

Larger sized dredges will not speed things up under all conditions, however. A five inch dredge will go into a lot of places that a six or eight inch dredge will not go—meaning fast water and shallow water. Also, bedrock can be so rough and irregular at times, that you cannot get an eight inch nozzle into the cracks and crevices far enough to recover a lot of the gold. When you run into this in shallow material, sometimes you can get more done with a smaller sized dredge. It also takes more visibility to work a larger dredge. And so there are times when an intermediate can move more volume than the larger dredges, and this is another reason—besides sampling—that it's a good idea to have one.

The secret to operating a thick walled, larger sized suction hose effectively, without wasting too much energy, is in positioning the dredge so that you do not need to do any more flexing of the hose than is necessary. You will have to move the dredge forward more often this way, but you will also get more done with far less energy. Otherwise you'll be fighting the hose most of the time—which can really wear you out.

One good way to position a smaller sized hose—like a five or six incher—is balanced over one thigh; so you can operate the nozzle with one hand, while tossing cobbles with the other, as shown in figu re 3-14. (next page)

FAST WATER

It's usually safer to dredge in shallow fast water, than it is to dredge in deep fast water--but it can still be very dangerous!

A lot of the time, it's necessary to wear more weight while dredging in fast water. Even if this is the case, the fewer belts you use, the better off you are if you need to get rid of them in a hurry. One of the main reasons for this, is when you are dredging with more than one weight belt on, they tend to shift and rotate around so that it's tougher to find the quick release catches--especially under emergency conditions.

You can get more weight onto a single belt if you use the 15 pound weights instead of the smaller sized ones. These tend to shift around the belt on their own; but this can be permanently prevented by drilling holes and bolting the weights onto the belt so they won't slide around, as shown in figure 3—15. (next page)

Figure 3-14: Balancing the suction hose over one leg so it can be operated with one hand while the other can be used to move cobbles.

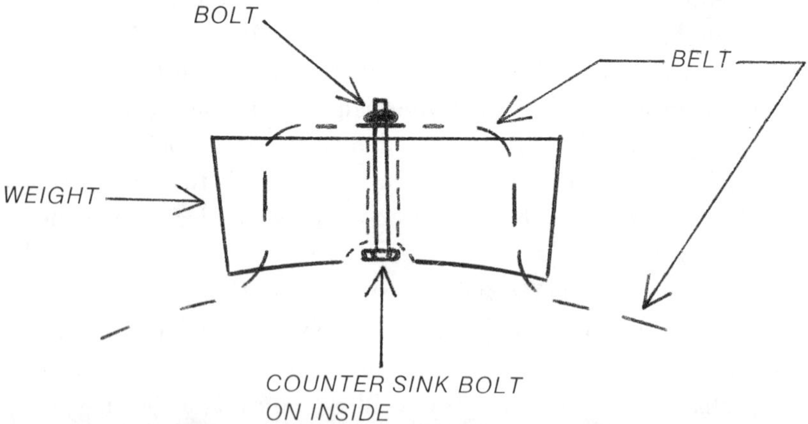

Figure 3-15: The larger sized lead weights can be bolted to the belt to keep them from sliding around.

Fast water dredging!

When I dredge in extremely fast water, I have found it best to not attach ANY lines to my body at all--not even my air line. It seems like it would be much easier to get back to the water's surface, or to work your way back in to the streambank, or to release your weight belt, if you are not dragging from any lines--especially an air line. Some situations may be different than this--although I have yet to see any. YOU WILL HAVE TO USE YOUR OWN JUDGMENT IN ALL SAFETY MATTERS! If, however, you

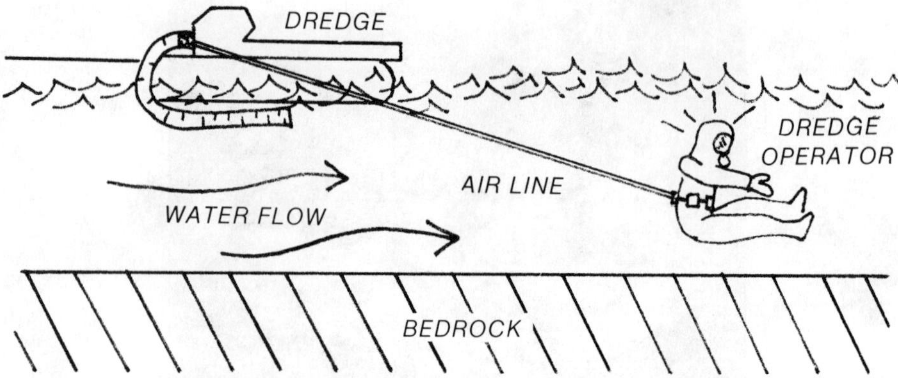

Figure 3-16: Attaching yourself to any lines in fast water can be doubly dangerous!

do end up dragging from your air line, it will kink and cut off your air supply, and hold you there helpless in the fast current--I'll bet it's not a very nice feeling! Dredging in very fast water can be extremely dangerous--and I do not recommend doing it unless you are really up to it and have some experience!

When dredging in swift water, your suction hose can become swept out of your hole and down river to drag from the power jet. When this happens, the easiest way of pulling the hose back forward again is by starting at the jet and working it forward, as shown in figure 3–17. This takes much less time and energy than going back to pull the nozzle forward.

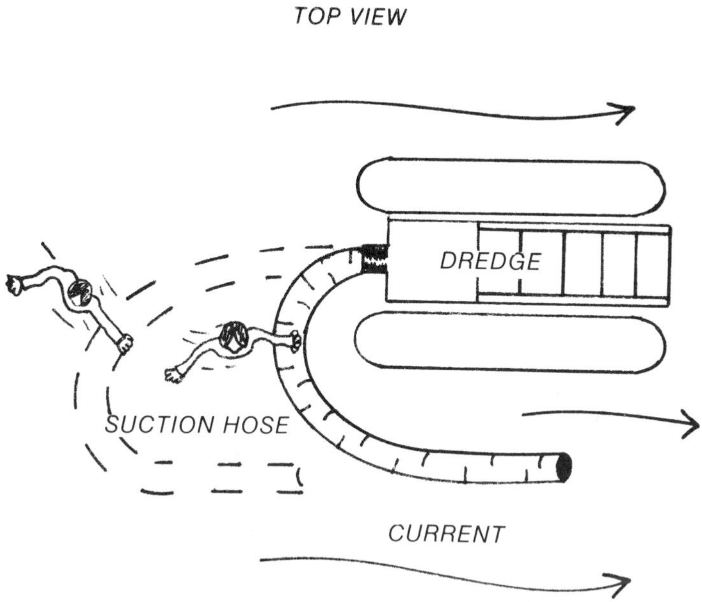

Figure 3-17: Working hose forward from the jet.

Dredging the fast water!

FAST WATER HARNESS

When dredging in moderately swift water, there is a harness you can easily rig up which will keep you from losing very much of your suction hose if it gets swept out of your hole by the current, and which also makes it much easier to quickly pull your dredge further upriver when you need to. The harness is rigged so the dredge tie off is located right above where you are working the nozzle. This way, when it's time to move the dredge forward, you can do so and pull the hose forward at the same time. (See figure 3–18.)

After you have set up your dredge in a new spot on the river and have started dredging again, it's a good idea to look back every once in awhile to make sure the dredge is riding well. This is especially needed when you are dredging in fast water.

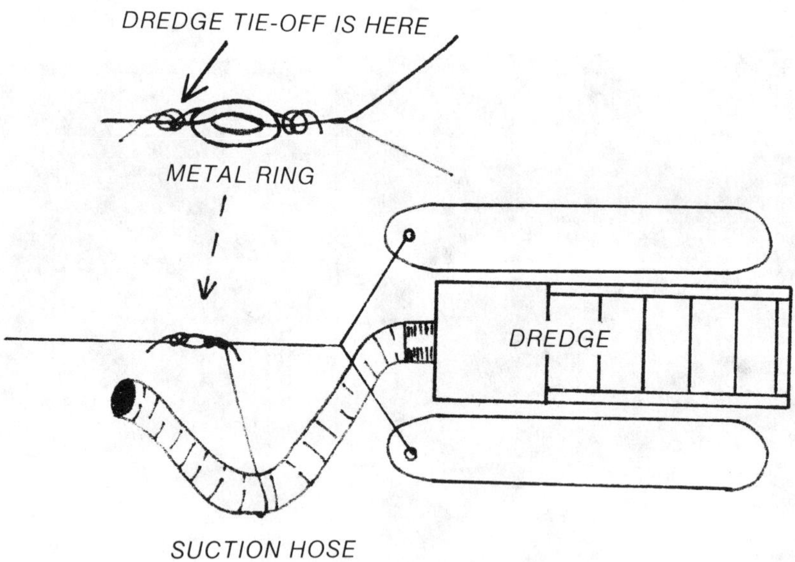

Figure 3-18: Fast water harness.

MAINTENANCE AND REPAIRS

Equipment failure can be one of the largest problems and expenses to a gold dredging operation.

Store bought dredges, for the most part, are dependable, have all of the major bugs worked out of them already, and will last quite a few seasons as long as the normal upkeep is done, and the minor repairs are handled when necessary.

Minor repairs, if not tended to immediately, can be the cause of larger and more expensive problems.

Unless you have seen it, you would not believe the amount of time and money that can be spent working out the bugs on a home built suction dredge! For that matter, you might not believe the amount of time and money that can be spent on dredging equipment which has not been kept up properly during a production operation!

Every single bit of cash spent needlessly on equipment, is cash and gold no longer in the pocket!

CHAPTER 4

DEALING WITH GOLD, CLAIMS, AND PARTNERS

Human relations can sometimes be a very difficult subject to deal with. Gold in itself is a highly explosive commodity--always has been; and so add the problem of human relations into the bargain, and you are potentially in for a lot of grief--or someone is--if it is not handled carefully.

When dealing with raw gold in any large amounts at all, you have to be careful about who you partner up with, and how you partner up--especially when you are dredging up large amounts of gold out of a river, or plan to do so.

If there is anything on this earth that is likely to hit a man where he really lives, paying quantities of gold is likely to be one of them. There is just something about the yellow metal which causes a person to admire it, and to want it. And there never does seem to be quite enough of it to completely fulfill ones needs. Gold is a very rich and beautiful substance to deal with. The more of it you have--and the more you get to looking at what gold you already have, the more of it you tend to want.

Perhaps gold does not strike everyone like this; but it certainly does strike most miners in this way. I have heard quite a few guys say that they are immune to all of this, saying they are just into dredging for the money and economic reasons. But I say that until you uncover a very rich deposit of your own, you really cannot be certain of your immunity, or how you will react. And even if you are one of the few who are free of the emotional impact from large quantities of gold, if you are going to deal with it, you better plan on the effect that it will have on others!

I once knew a very knowledgeable guy who had spent a considerable amount of time studying the characteristics of gold--along with other things. One of his conclusions was that each different kind of matter (the basic elements) emits a different type of wavelength, and that the wavelength put off by gold is a wavelength of beauty--in the aesthetic band. Whatever! That gold is a very beautiful substance is not in question, and that people will extend themselves to great lengths to aquire gold has certainly been proven out in history.

So in setting up the various dealings which you will have with partners and claim owners, you need to be extra careful to make deals in which all parties are agreeable to--and in which all parties will stick to. Therefore, all agreements should be made very clear--and in writing if possible, and ALWAYS BEFORE THE GOLD IS DREDGED UP!

It's really not enough to say: "Lets not worry about how we're going to split up the gold; we'll figure that out if and when we find it." This might work alright on other things, but with gold it's very poor organization. Arrangements like this can very easily end up in dispute once the gold is in hand--or as soon as it is in sight, for that matter.

Additionally, any agreements that you do make in a gold mining venture, should be made to suit the amounts of gold that you are going after; the reason being, that if you make a deal which seems to be fair with everyone concerned based on dredging a barely acceptable paystreak, and you suddenly are into a very rich paystreak--making thousands/day, the deal may no longer be at all fair to everyone involved. But to try and bring this up and put together a new deal once large quantities of gold have been brought into play--and to get everyones agreement on a new deal--is almost impossible.

I learned this the hard way with an experience of my own.

I had been dredging in a certain general area for quite some time, and had really put out a lot of energy to get familiar with it. Through dredging many many sample holes, I was working my way closer to a very large and excellent paying stringer--but I didn't know it at the time, because I had not found it yet. Man, I had gone through some pretty lean times prior to that; it was winter dredging--cold water, storms, etc., and I was trying to learn how to dredge and how to find stringers. I was just starting to get pretty good at it--but I didn't know that either, and that's when an old friend of mine asked to come out and dredge with me for a few weeks. Well, he was a good friend, and his company sounded pretty good, so I let him come on.

We worked out a 48/42% deal--which seemed about right; because at the time I was not into any paying quantities of gold; and it seemed like he ought to have a fair sized split of anything that we found together--but that I should receive a larger share, because it was my operation and he was just visiting.

Well you will never guess: We hit that rich stringer on the first hole of the first day that he arrived! It was not the first stringer that I had been into, but it certainly was by far the largest and best paying--even after our split.

But the thing about it is that I was planning to put that sample hole there as my next project anyway. I had been working right into that stringer for months in my data collecting, hard work, and sampling activities. (That and half dozen others for that matter--which I didn't know then either.) Man, were we into the gold--it was incredible!

And he received his 42% split; because that was our agreement--yet I had put a year or more of work into learning how to dredge and find that deposit; and that was the gold which should have paid me off for all the hard times I'd stuck through up until that point. Because of the deal I made with my friend, I ended up giving almost half of it away to someone who took no part in the hard work of finding it. Sure, he helped clean it up; but cleaning up a rich stringer is the fun and easy part. The amount of time and energy it takes to get familiar with an area and learn how to find stringers is worth a lot more. But this is something that my buddy would never have been able to fully appreciate--because he never had to go through it. To him it was just good fortune, and to have tried to explain differently would have caused hard feelings. Besides, a deal is a deal; so he received his full split from the entire stringer and went away very happy. Me? I learned this lesson well: *MAKE YOUR PARTNERSHIP DEALS BASED ON THE AMOUNTS OF GOLD THAT YOU ARE GOING AFTER--AND THE CORRECT EXCHANGE THAT SHOULD GO TO EACH PERSON WHEN THAT MUCH IS BEING RECOVERED!*

It's very easy, by the way, to make a generalized deal that you are not particularly happy about, or in which all the details are not fully worked out, before you get into the gold. This is easy to do if you are not sure that you are going to find gold in paying amounts anyway. Then, after the gold is discovered, you wish like mad that you spent more time doing it right in the first place! I've seen this kind of thing happen time and time again. It's so easy to talk glibly with your partners about how you won't have any trouble dealing with the gold, when "the gold" is still an abstract concept that you are uncertain will ever become a reality. It's another thing entirely when you are pulling it out of the river by the fistful, and you feel entitled to so much because of your contribution; your partners feel the same way because of their investments; and there is not enough gold to go around on that kind of split--especially the nuggets! You can lose good friends over the smallest of disagreements when it comes to gold! No, it's better to figure it all out before the gold gets onto the scene. For that matter, when dealing with LARGE amounts of gold, it's tough enough to prevent dissention within a group even when the best and most detailed of deals are made with the most ethical of partners! This is just the way it is with gold.

So, it's always a good idea to plan on the gold being there--since you are going after it anyway, and to try and work all the details--and any disagreements--BEFORE you even start looking. Quite a few things need to be discussed, like: exactly how and when the gold is going to be split up once you get into it; how the nuggets are going to be split up; who--if anyone else, is going to be allowed to help clean up the stringer; who--if anyone else, is even going to be allowed to know about the stringer before it's cleaned up; how much time is going to be put into production; how much gold is to be sold to handle expenses and repairs; where the gold is to be kept before it is split up, and etc.

In the case of partnerships in gold deals, I also feel that not only should all the agreements be thoroughly communicated about with everyone concerned, but the agreements should also be put into writing with everyone signing and receiving a copy. This not only forces all members to discuss all the important parts; but it also provides a written record of the agreement, should any disagreements or questions arise later. It's a nice feeling to know exactly how things are going to go once you start uncovering the gold.

Since conditions do change, sometimes agreements need to be changed--which is fine as long as everyone agrees. When this happens, the written agreement needs to also be modified as soon as possible. Otherwise, it is useless from there on.

SPLITTING UP GOLD

I worked out a pretty good method of splitting up the gold with my partners on an Alaskan venture last summer, which worked out very well; and you might be able to make use of the method in your own operations. Since it was a group venture, all recovered gold went into a group collection jar. We would pull out any and all pieces of gold that seemed of extra value, and put those into a separate group jar. These we saved until the end of the venture before being split up. The rest of the gold--meaning the fines and flakes--were split up on a regular basis. This we would screen through window screen, and an equal share by weight of each classification would go to each partner. If a partner was to receive a slightly higher percentage, he would receive it in both classifications.

At the end of the venture when it was time to split up the nuggets and jewelry gold, we picked out the most extraordinary nuggets, put them aside, and made equal piles of all the rest--equal by weight, and equal by size and apparent value classification of the nuggets involved. When everyone was satisfied that the piles were all equal in value, we drew cards; and whoever had the highest card, chose the first pile; the second highest card chose the second pile, and so on. If a partner had a slightly higher percentage of gold coming to him, he would receive the proper amount out of fines and flakes which were set aside for this purpose.

Then we took the most extraordinary nuggets that were left for last, and bid them off one by one--just like in an auction. Everyone had plenty of chance to bid up the amount of fine and flake sized gold they were willing to pay for each nugget. Each nugget went to whoever was willing to pay the most; and the gold payments went into the group jar, to eventually be split up properly amongst us--after all of the final trip expenses were paid.

This system worked out very well, with no hard feelings, even though we were dealing with a lot of nuggets.

DEALING WITH CLAIMS

Dealings with claim owners should also be done in detail BEFORE a dredger starts sampling out different mining properties. It is for the gold dredger's protection that I bring this up. I've seen it happen quite a few times, where a dredger makes a generalized agreement with a claim owner, whereby it's OK for the dredger to prospect around on the claim, and that they will make a "fair deal" if pay dirt is found. A deal like this might seem fine and dandy before gold is found; but things always look entirely different to everyone involved once a good paystreak is located on the claim.

Before a paystreak is found, the claim does not necessarily carry any great value from all appearances; because you do not know for sure if there is even any pay dirt on the property at all. This is the best bargaining position for the dredger, and a ten percent deal is quite standard when this is the situation--that's 90% to the dredger.

Once a rich paystreak is found, if a specific percentage deal has not already been made, the claim suddenly becomes very valuable; and the claim owner is in a position to be asking for just about any percentage that he chooses. His position will now be that a treasure has been located on his claim, and all that's needed is for someone to come along and dredge it up-which is easy! So if a general deal is made "to be worked out in detail if paydirt is found," and the miner does locate a stringer, the claim owner might feel entitled to 50% of the find--or more. "After all, it's rich paydirt on my claim, and all you have to do is dredge it up: 50%--take it or leave it!" You can really get the short end on such deals; because the claim owner does have the upper legal hand; he can throw you off!

So it's not at all a good idea to strengthen a claim owner's bargaining position by finding a stringer on his claim before you finalize a percentage deal. The final deal should be made before sampling is started, with the claim's worth taken at face value at that time.

Just like in partnership deals, the arrangement with a claim owner should be thoroughly discussed: Is the ten percent going to be of recovered gold weight? Or, is it going to be ten percent of the complete range of nuggets, flakes, and fines? Is the split to be done on a daily basis? A Weekly basis?

Monthly? Or when you leave the claim? Does the owner want to see the gold split up? These are the types of things that need to be agreed upon during the early stages.

Another thing that should always be discussed with a claim owner is the time frame. If a dredger goes onto a claim on a ten percent deal and then finds a stringer, how does he know that the owner is not going to suddenly decide that the dredger's time on the claim is up? One way is to discuss it at the very beginning: "If I find a stringer on your claim, I am to be given the time to finish cleaning it up, right?" I've seen even some of the most experienced pros goof up on this point, and I've done so myself a few times; and let me tell you--it hurts!

If you are not receiving an exclusive lease right to the claim, another important thing to discuss with the claim owner during the negotiation stages, is your exclusive right to any stringer you yourself locate on his claim. This is making sure that no-one else is going to be allowed to move in on your stringers, or that no-one else is to be allowed to move onto the claim up in front of you and cloud you out with their tailings. This is something to consider, and discuss--because it happens.

It's not always necessary--or possible--to get such agreements in writing with claim owners, but I feel it's good to do so when the guy you are dealing with looks a little shady, and the formalness of your request is not going to blow the deal for you. Some guys simply do not like to commit themselves to anything in writing, and you cannot push it without causing some kind of upset. This doesn't necessarily mean that these people are lacking in integrity. Actually, there are still many people around who may consider that there is something wrong with your integrity if a hand shake is not good enough to seal a deal! Of course, the more you put at stake on a single claim, the more important it is to have a written contract.

In any case, written or not, it's best to discuss all aspects of the deal ahead of time, and come to the agreements that will be followed. These are the rules of the game; and it's good to discuss them long enough, and thoroughly enough, to insure that both sides understand all the details--and are in full agreement with them.

PERCENTAGE DEALS

For the most part, at least in my neck of the woods, the claim owner usually receives ten percent of what a dredging operation recovers off of his claim. This is a claim owner's standard cut—unless the section of river has already proven to be rich, at which time the claim owner may be entitled to 15% of the find.

Personally, I feel that 20% to a claim owner is too high if you have to locate the pay dirt yourself. If he can tell you right where it has already been located, and it is just a matter of you going in there and dredging it up, maybe 20% or even a higher percentage might be OK—depending on how rich the location is.

You are going to pay the operating expenses, repairs, and living expenses; you are paying for the equipment, are doing all the hard work, are taking all the chances; and if you need to find the gold too—on the chance that it might not even be there in the first place, twenty percent is too much of a demand by a claim owner. But some will ask, so be prepared for it.

One thing to keep strongly in mind while negotiating a claim deal, is that the larger percentage you agree to pay a claim owner, the less percentage that you will end up with as a profit after all the expenses are paid. This means that the more you agree to pay, the better the claim has to be in order for you to make a profit. There is a lot of ground around that might meet your minimum requirements on a 10% deal, which might not meet those same requirements when you have to pay 20% to the claim owner.

For a claim owner to ask, without a good and visible reason, for some unusually high percentage, he is inviting the dredgers to high-grade on him and he most likely will end up receiving less than what would be reasonable. Personally, I have my own philosophy about high-grading—which I'll be getting to shortly; I just don't do it, and with good reason. But I have talked with a lot of miners, and have found that for the most part, unreasonable claim owners usually get treated unreasonably!

KEEPING QUIET

One good way to get yourself into all kinds of trouble, is to start yelling and screaming out like a mad man when you have just made some kind of find—even it it's just a little find, or even if you are just goofing off! Sound carries far; and those people who overhear you and your excitement, imagine gold finds and bonanzas probably far better than what you are into—claim owners included. If you find a little gold and make some noise about it, word gets around; and the claim owner might think you are cheating him come royalty time—no matter how much you give him. If you find something nice and keep quiet about it, it will be a pleasant surprise to the claim owner come royalty time when he gets more than usual—get the idea? Don't get me wrong, I'm not saying to withhold information from the owner; I'm saying to prevent allowing your excitement to create false information which can thereafter get you into trouble.

And there is something else about this that you should be aware of: I do not know exactly why it is, but when a guy suddenly starts to do really good at something, there almost always seems to be someone else around who cannot stand to see it, and who wants to give him trouble. This can be especially true when dealing with gold.

When you are into a rich paying stringer, it's best to keep quiet about it until after you have finished cleaning it up. Then you can make all the noise you want to. But while you are into it, if you don't tell anyone who doesn't need to know, you are more likely to keep the dogs from biting at your heels. The dogs always like to hear that you are going broke, and that is always the way it looks with my personal operation when I'm into something good!

INTEGRITY AND GOLD

I personally *NEVER* agree to a percentage deal that I do not fully intend to pay in the end. I feel that my personal strength and well-being as a person are worth far more than all the gold in the river—or all the nice things in the world that can be bought with it. I know there are quite a few people around that do not feel this way; but there are also a lot of people around who do—and these are the ones who make life worthwhile.

There probably isn't anyone around who has not cheated in one game or another at some point in his life. It is a very degrading activity for a person to be—let's say, the banker in a game of Monopoly, playing with his friends, and stealing money from the bank in order to strengthen his position in the game when the other players are not looking. The person himself feels this degragation. Indeed, anyone who can play a game with his friends, and cheat at it to win, and not feel some degree of guilt or remorseness afterwards, is pretty far beyond hope, and probably is not much of a friend at all—to anyone.

Any game or activity at all will be found to follow an agreed upon set of rules, for which there is a certain amount of freedom to act within. These are the same rules that are being formed up when a business agreement is made with partners or a claim owner.

No-one can truly win a game by breaking the rules which make up that game. To knowingly break the rules of a game in an effort to win, is to admit that the game—by the rules—is above ones ability to play. When a person needs to cheat or break the agreements which he has made in a game, in actuality, he has already lost! No game carries a stiffer penalty, and greater personal misery, for having lost because of cheating or unethical behavior, than the game of life.

It has long been known that when a person finds himself needing to make justifications in order to make some action alright to do—or have done, then the rightness of the action must be in question in that person's own estimation—or he would not need to be making justifications about it, right? I'm sure that everyone has experienced this.

To be able to see different degrees of rightness from wrongness, greatly determines a person's ability to differentiate between other things as well. A person's ability, and his sanity, largely depends upon his capacity to tell the differences between things.

A floor cleaner can tell the difference between a dirty floor and a clean floor. If he is good, he can tell the difference between a clean floor and a sparkling floor; and by being able to differentiate in this way, he'll be able to learn exactly how to keep the floor sparkling—and that's what makes him a good floor cleaner.

A gold dredger should be able to tell the difference between hard packed material, and slightly less hard packed material. Seeing the differance can mean finding a rich stringer or missing it by a few feet. There are thousands of tiny signs and things to see and know to succeed well in this field, and the good gold dredger sees many of these little differences and is able to follow up on the right ones. This is

ABILITY that I'm talking about, and it depends entirely upon a person's capacity to tell differences.

A person's sanity, personal well-being, happiness, and personal strength, greatly depends on his ability to tell right actions from wrong actions—and to follow through with the right ones. When a person sees that some action which he is planning to take, or some action which he has already taken is wrong: that is part of his own basic self which is able to see these things; and it's his own basic sanity and sense of rightness and decency which is speaking out against the wrong action. When he has managed to justify the action off as a "right action," and has surpressed his ability to see the wrongness of the action: he has succeeded in shutting off part of his ability to see differences, and his ability to follow through; and he has managed to shut off a portion of his basic self and his future happiness. Eventually, if a person continues on in this manner, he can shut off so much of his own basic decency, that he can lose track of who he *REALLY* is!

Of course, a person can grow, can become more able, can gain more awareness, happiness, personal strength, and well-being, and can regain more truth in his life by taking the opposite route of paying attention to rightnesses—and following through on them.

A person can cheat his way through life in order to succeed financially—and many people probably do; but do such people really succeed? Are they really happy? They can sometimes put on a good show of happiness for other people to see; but if you really look into the personal lives of such people, you will find that they are heavily burdened with misery, discontent, and sickness; and you'll find that their close interpersonal relationships are not satisfactory to them; and also find that they are wondering why life has been so hard on them . . . How could any person be truly happy knowing that his success is a result of his stealing it from those who worked for it? What good does financial success do if you have to cheat to acquire it, and therefore cannot achieve real happiness to enjoy it?

It would seem that real happiness could be obtained by playing the game fairly, and that the game could be won by hard work and application of one's abilities in an ethical manner.

A person who can make an agreement with another person, knowing full well that he has no intention of following through with his part of it, will be found to be missing some of the more important things in life, like: honest friendships, the sense of belonging, caring, loving, nice feelings towards others, and feeling good about himself.

When you tell a claim owner that you are going to give him a 25% royalty, and you only give him 3% in the end, while leading him to believe that you are giving him 25%—for whatever the reason (justification), it will screw your head up, and there is really no getting around it.

This is not a book on philosophy, or on religious beliefs; it's a book on how to improve your gold mining skills. If I did not know that this section on integrity was positively an important and necessary part of the whole, I would never include it in this advanced manual. If you want to improve upon your ability to find and recover lots of gold, and get real good at it, work also on improving your ability to tell differences. If you want to improve upon your ability to tell differences, work on improving your personal integrity.

Interestingly enough, the vast majority of continuously successful gold dredgers that I have made contact with, are honest, hard working people who are serious about following through with agreed upon royalty payments—almost to the point of religious ritual. But it is the same ability to see and follow through on right actions which is used in tracing down and recovering gold deposits; so this is no great surprise.

People who are good at gold dredging do not need to high-grade in order to succeed and do well at it. People who are not yet good at gold dredging need to improve their skills through lots of study, hard work, and fair dealings. There lies the road. High-grading and other forms of unethical dealing, is a freight train going in the opposite direction!

This is all something that you need to look at for yourself, kick around in your own mind, and figure out how you are going to make your stand; because if there is anything in this world that will put your personal integrity to the severe test—it's gold! Gold is a highly explosive commodity to handle. It can be very hard to give away once it's in your hands. Gold is especially hard to give away when you do not feel that the recipient deserves to have what it is coming to him—even if you have agreed to it beforehand. In such a case, with gold, sometimes it's tempting to not follow through on an earlier agreement. You can find yourself coming up with all kinds of reasons (justifications) as to why you do not need to follow through with the earlier agreements. Justifications can be made to sound very reasonable. But if they are that reasonable, why not take them up with the claim owner?? A very tempting commodity, gold is; and you need to be careful to not let it ruin your life!

This is why I always try and work out a complete deal before I go onto a claim. There is much to lose by not doing so. If I cannot come to acceptable terms with a claim owner, it's nice to know ahead of time. There is no scarcity of gold bearing river around. I stay away from bad deals; this way I never put myself into a position of putting my personal well-being at stake.

I also stay away from making deals with bad people. Such people have a severly contagious disease. Take Joe Blow for example, who is doing pretty well in life and making progress towards his goals. Then one day he takes up a partnership with Pete Shmuck—who is obviously a shady character. Let's say that Joe has taken a pretty strong stand on his morals, and so doesn't get pulled down into unethical activities by

Pete Shmuck. One day, however, Shmuck finally ends up pulling the big con job on Joe Blow—or on Joe's associates, which is even worse! Joe gets mad and now feels the need to retaliate, or to get even, or to stop Shmuck, or if nothing else-to make things right again with his associates-if that's even possible. But before Joe teamed up with Shmuck, he was doing fine and getting along well in life. See what I mean? Dishonest people are sick. They are losers. They are dishonest because they are not strong enough to do things the right way. So they really do not have much positive force to offer the hard working, honest man.

It is not so much the things that have been done to one which ruins his or her life. People can contend with hardships—and become stronger because of them. It's the harmful actions against others which are hard to live with, even if they are well justified. Everyone has a natural right to protect himself and his interests. The best protection is to not connect up with the bad elements. If you cannot keep someone from committing harmful actions, you are better off to disconnect entirely from that person. Even rightful and just retaliation in today's world is likely to get you into jail! So there is very little winning in the "get-even" game—better to disconnect.

The reason why I've gone into all this is because gold can really bring out the harmful intentions in bad people. So you have to watch who you partner up with—and who you deal with too. There is s saying—and it's a good one: "If you want to fly like an eagle, quit playing with turkeys!"

GROUP DREDGING

After having observed many different kinds and sizes of dredging operations, and in having taken part in different sized group operations myself, I have pretty much decided that gold dredging, for the most part, is better done as an individual activity, or as a very small group activity—like two guys, instead of as a larger group activity. There are exceptions to this I'm sure; but if a dredger knows how to find and recover his own deposits, and has his own equipment, he will end up with more gold in his pocket in the long run if he does it on his own—rather than attempt to make gold by keeping other more inexperienced dredgers continuously into deposits.

A good prospector should be able to keep himself into gold rather consistantly. A stringer on any fairly large sized body of running water (river) is likely to run 15 ounces, 40 ounces, 80 ounces, or even more. It may take a dredger two weeks to clean it up, four weeks, or maybe even the entire summer. When an individual gets himself into a stringer, he is going to be busy for awhile making good gold—and what he recovers is his.

When an individual is trying to keep four dredges into gold, he is only going to receive about a quarter of the total recovery; and the stringer is going to disappear four times as fast. If some of the other dredgers are inexperienced, it's likely that some of the stringer is going to be missed and left behind. This means that the guy has to find more than four times as much gold to make the same amount himself in the long run.

By far, the hardest part (and most valuable) of gold dredging is in finding the stringers. It's not hard at all to dig up a treasure once it is located—but what about finding it in the first place? And it is similar to this in a group dredging operation.

An individual, by himself, may only need to find one or two stringers during an entire season to keep himself plenty occupied. The rest of the time, he will be busy dredging them up and making good gold. If the same individual needs to keep a four man operation into gold, he might have to locate ten or more stringers to make the same amount of gold himself. That would be enough to keep him busy alone for the next five years or so! And even if he does it, he will have spent a lot of his time sampling with one dredge, which means only three dredges will be into gold—on a four way split. Because of the nature of stringers, it's very likely there will be many days when two or more dredges will not be into paydirt, or when a dredge is broken down, or when an operator is sick or has personal problems to handle, etc. So in the long run, the individual dredger will probably make less in a group dredging activity, even if he is good enough to find that many stringers.

There are other things that go along with this; like an inexperienced dredger will not be able to move near the volume of material that an experienced dredger can. Also, whereas an experienced dredger can easily see and follow where a stringer goes in the streambed, an inexperienced dredger, if he is not supervised closely, is likely to be dredging up the low-grade material and dumping his tailings on top of the stringer. But you cannot supervise and go off sampling at the same time; so there is more lost production and less gold being recovered.

There's more too: A person who's income depends on the over-all production of a group may not have as much incentive to work his tail off as he would if his income depended entirely upon his own production—especially if the dredging conditions are really tough. There may be exceptions to this, but I have not seen very many of them in gold dredging. When the gold that ends up in your own sluice box belongs to you, you will find yourself working harder and longer hours to recover more gold, find more stringers, and learn how to do it better. I've never seen a group dredging operation succeed yet that paid it's dredge operators by wages, instead of a percentage of the gold. And I 've only seen a few large scale dredging operations (5 or more guys) work out that were based on a split of the gold found amongst the dredgers involved. Most of these were located in areas where large sections of very rich pay dirt were already located, and all that was necessary was for the dredgers to get in line and clean up all the material off of bedrock—no further sampling was needed. Conditions like this are not so plentiful that you can depend on finding lots of them readily.

Please do not get me wrong on this; I've nothing at all against organized group activities, I'm in favor of them under most conditions. I just have not seen them work nearly as well as individual activity in gold dredging. If you are going into it mainly for the purpose of recovering the gold—and recovering as much of it as you ethically can for your efforts, then you are probably better off to do it as an individual activity by yourself, or in conjunction with another gold dredger in a limited partnership.

PARTNERSHIPS

There are several good reasons to team up with one other person in a gold dredging operation; one good reason is safety. A dredging accident is less likely to be fatal if there are two guys on the scene keeping an eye on each other. A buddy is also able to offer a huge amount of assistance when it's time to move your gear around during the sampling stages. Still another good reason to have a partner—which is probably the best reason, is the moral support that two guys can give each other when the going gets tough: which can be a lot of the time during sampling, and when you are just learning to sample.

It is true that two guys are likely to finish off a stringer twice as fast, with only half as much going to each person; but it's also true that two guys tend to spur each other on and get more done—as long as both are hard workers. It can be much easier to locate the stringers when two guys are sampling and seeing it through together. And one other thing: In gold dredging, there can be a lot of hard times; but there can also be a lot of amazingly good times—which carry an incredible emotional impact; like when a very rich deposit has been located, and is being dredged up. These are experiences which could never be wholly communicated and understood by most others; so it's nice to go through them with someone else.

Most stringer deposits can be dredged up with two dredges working side by side. To try and mine stringers with more than two dredges at once usually becomes confusing. A lot of the time, there simply is not room for more than two dredges—especially when the water is deep or is moving fast. So it would seem that under the majority of conditions, two man teams are better than larger sized group dredging activities—and by that I mean that the individual dredgers are likely to end up making more gold.

PARTNERSHIP DEALS

There are several ways that two guys can team up in dredging: They can form a full partnership and find, recover, and split up all the gold equally between themselves; they can form a limited partnership and help each other to locate the deposits and clean them up together, with each person keeping what he himself dredges up; or, they can form an even more limited partnership, and agree to help each other move the gear around when needed—with each person finding and recovering his own deposits. Each of these different arrangements has its own advantages and disadvantages, and each can be most optimally used under a different set of circumstances. It's good to be able to shift from one of these to another, depending on what the situation is.

Teaming up with one other person on a full 50/50% partnership basis is done best when both dredgers are of relatively equal dredging and sampling ability, when both are willing to put equal amounts of time and energy into the operation, and when both have about an equal share of investment into it too. Just by nature of the way gold is—and by the way it affects people, if the gold being produced by the operation is going to be equally split up, then the amounts of input into the operation need to be reasonably equal too. Otherwise, dissension is likely to occur between the partners. And the more gold that is being recovered, the more this equation tends to show itself. This can happen in varying amounts, depending on the nature of the people involved, how much gold is being recovered, and how far out the exchange of time and energy input has gone between the partners.

For an example of this, lets say that two guys form a full 50/50% partnership, and one of them puts in most of the equipment, and most of the hard physical work. The other guy puts up an equal amount of operating expenses, but is really just along for the ride because his friend needs a partner. When the stringer is found, the first guy also puts out most of the time, energy, and driving force to recover the gold. In this case, a 50/50% split of the gold might not be fair; and if it isn't, it will be felt by the first guy–who did most of the input, and by the second guy–who is receiving more than he rightfully deserves. And if they are dealing with large quantities of gold, it is going to be felt a lot more than if they are dealing with just a few ounces. This is one reason why full partnerships are done best with another person of similar dredging ability and willingness to work. It's also a reason why it's good for you to be willing to switch from one kind of deal to another–as long as it's agreed upon by everyone involved.

I was dredging on a 50/50% deal with another dredger in a moderately rich stringer not too long ago. We each had our dredges in the stringer side by side, but I was recovering two to three times more gold every day than he was. This can happen once in a while in a stringer when one guy will uncover a richer section of paydirt, but this was happening every single day!

The water was very cold, and it was towards the end of a long and hard-working, successful season. We were both kind of burnt-out, but I had set a quota for myself to put a certain number of additional ounces into my poke before knocking off for the season. The gold was there; I just needed to dredge it up. My buddy needed the gold too, but he was more burnt-out than I was; and the only reason why he had not called it quits, he said, was because I hadn't.

One day we sat down and I told him that I only wanted so many more ounces before ending off; and that I wanted to go after them on my own–keeping what I recovered, and allowing him to keep what he recovered–if he chose to dredge at all. Since the stringer was just as rich on his side as it was on mine, and since he was an ethical man, he could see the fairness in what I wanted to do. So he agreed, and we started dredging side by side, each of us keeping what gold we recovered. Well, you will never guess: I started recovering a little more gold than I already was, and my buddy started recovering about three times as much gold as he was before! The stringer did not get any better over there; we just changed a working agreement. This is fruit for thought.

GRUBSTAKING

There are various ways that different guys can contribute to a dredging operation to keep the input relatively equal. One guy may have the equipment and the experience, while another guy might have the money to finance the operation. These kinds of deals can be made to work when both parties are equally willing to put time and physical energy into the operation.

Those kinds of deals where one partner is going to put up the money and end up with half the gold, while the other guy does all the work, do not work too well. You really have to be careful to not team up with somebody who isn't going to put in his equal share of dredging. Since you can really only put in so many hours a day at dredging, there is still plenty of time left over to clean-up, cook, repair gear, and so on. So there is no good reason to give an equal share of the recovery to someone else to do these things for you; and if you do, you will be feeling it out of your own pocket.

I have seen grubstake deals work out when the percentage to the silent financier was kept down to about ten percent or less of the gross find. Unless it means not going at all, you should try and avoid having to give any more royalties away than you positively have to. As mentioned earlier, the larger a cut you have to give away, the more pay dirt you have to pass up that would otherwise be acceptable dredging.

Unless some very rich area has already been located and just needs to be dredged up, a grubstaker—and claim owner—will make more gold in the long run if the amount of gold the dredger needs to pay out is kept down to a minimum reasonable percentage of the recovery. This way, the dredger is allowed to continue at it with success, instead of going broke. I met up with a guy last summer in Alaska who had agreed to pay a grubstaker 50% of his gross recovery; because the grubstaker had sent along enough groceries to last the dredger all season! Now let's see, that's 50% to the grubstaker, 10% to the claim owner, and the dredger still had to pay his own travel expenses, operating expenses, repairs, and etc. How was he going to make any money for himself? He didn't—and neither did the grubstaker! And the worst of it is that he was into a very nice stringer—which he had to leave behind for someone else to dredge up.

OTHER DEALS

I've seen other deals work out also, where one guy might receive a higher percentage of the find because he has put more into the operation than the other. This works, and is often necessary to do; but regardless of the agreement—and how right it is, when split-up time comes, and one guy's pile is sizeably larger than the other's pile, both guys are likely to feel a little strange about it in some way. This comes into play more when large quantities of gold are being recovered—making the difference in piles even more noticeable. This is the way it is with gold, so it's best to plan on it.

OVERMANNING AN OPERATION

One poor way to set up a full partnership, is by teaming up with someone who does not have an additional dredge to operate. One dredge can only process so much material in a single day. Two dredges of similar size, co-ordinated properly, ought to be able to double what one dredge can do by itself—as long as both operators are hard workers. If you take on a partner without also having a second dredge for him to operate, you are likely to make considerably less gold than you would if you work alone.

There are exceptions to this, like when dredging in deep material where there are a lot of cobbles to be moved—enough to keep a helper at it full time without him getting into your way. Another exception is when operating a large dredge, when it's a two handed job to move the nozzle around, leaving room for a helper to move the cobbles and boulders.

When two guys are put to work on the same dredge, it is necessary for them to be able to move nearly twice as much volume of streambed material through the dredge, and recover twice the amount of gold. Otherwise, the system becomes inefficient and they would be far better off to operate two separate dredges, or take shifts on the same dredge.

In theory, you might figure that if one man can move a certain volume of material with a dredge, then two men working together on the same dredge ought to be able to move twice as much material—or more; and if conditions are right for it, as mentioned above, this might be true. But if there is not enough work to keep a second man busy full out, without him cutting into the nozzle-man's production and slowing him down, then it's not a good idea to put a second man there. What do I mean by cutting into the nozzle-man's production? When a second man runs out of productive activity to keep busy, he's likely to start moving cobbles that are not ready to be moved. In doing so, streambed silt will be released into the water; and in not being sucked up, it will cloud out the hole—forcing the nozzle-man to wait until the water clears up enough again that he can see that he is sucking into the nozzle.

To give you an example of an overmanned (and therefore underpaid) production dredging operation, I teamed up with two other guys one time in a sampling activity to locate a stringer. We found a very nice one, which was sitting underneath about 2½-feet of material. The stringer was about 45 feet wide; so there was no problem in placing our two dredges side by side to clean it up.

We were each operating five inch sized dredges. I was running mine; the two of them teamed up on theirs; and we agreed that we each would keep what we recovered separately.

The material was red packed, and so it was necessary to take apart the streambed in an orderly method—sucking the silt and material from the same spot that you were moving cobbles. Otherwise, that part of the hole would be entirely clouded by silt, and you would have to wait for a few minutes before it would clear up enough that you could see what you were doing again. But this is standard operating procedure in most dredging situations.

The hole we were dredging was so wide, that if I clouded myself out, their side would still remain clear enough that they could continue on dredging, and vice versa.

Since we agreed that each would keep what we separately recovered, and agreed that each could spend as many hours at it as we wished, the game was to produce like mad to dredge up as much gold as possible before the other guys recovered it. Well those were some hard working guys and they surely kept me moving all the time–me knowing that they were dredging up all that gold as fast as they were. They had teamed up on their dredge and were going at it hot and heavy in the other side of the hole. But at the end of each day my tailings pile was at least twice the size of theirs, and I usually had twice as much gold too.

In wonderment at this, I stopped one day to have a look at what they were doing; and sure enough, they just kept clouding each other out. In 2 1/2-feet of material, there just was not enough work to keep two guys effectively busy on one dredge. I then went over and watched myself dredge, to see what a second man could do to help me get material through the nozzle–and there wasn't much; I was pouring it through about as fast as it could be done most of the time. The best bet for those guys would have been to split the day into two shifts: One starting at daybreak, and the other one ending at dark. In that way, each one of them would have ended up dredging about the same amount of gold that I dredged. As it was, they probably would have doubled their gold intake if one of them did not dredge at all! This is something to keep well in mind when setting up a dredging operation.

TEAMING UP ON A DREDGE

If two guys are going to effectively operate a dredge together, they will need to devise a system of co-ordinating dredging activities, so they can stay out of each others way and speed up production. There are various ways of going about this. I've picked up a few good points here and there in my own experience, which I'll include here for you to make use of.

I have always found that the nozzle-man (guy who is operating the nozzle) should be in charge of taking apart the hole. The other man ("rock-man") should act as his assistant and move rocks out of his way as needed to help speed up production. Otherwise, if the rock-man takes the lead and starts moving cobbles as he wishes, the nozzle-man will need to follow him around–just to keep the hole from being clouded out. But in this way, the nozzle-man will never really get a chance to get started at taking the hole effectively apart before the rock-man starts clouding out some other part of the hole again. Plus, the nozzle-man is the one who needs to be looking at where the gold is coming from so he can follow and recover all of the stringer. The nozzle-man should be allowed to take the hole apart in a safe, orderly manner, and concentrate on sucking as much material through the nozzle as he can; while the rock-man should concentrate entirely on moving whatever cobble that is next in the nozzle-man's way. He should do this as fast as he can, to allow the nozzle-man freedom to suck up material with nothing in his way to slow him down. This will increase production.

When I take on someone else to work with me, I usually start him off by pointing to each cobble which is the next one for him to move. Pretty soon he will catch on and start doing it correctly on his own initiative; because it's really pretty obvious what needs to be done once you get the hang of it. Then if we switch off and my partner takes over the nozzle, I'll follow his lead and move cobbles out of his way.

If you are doubled up on a dredge in rather shallow material, and there really is not enough work for two guys at once, you can try making your hole as wide as it will pay; and then dredge on one side, while the rock man moves all of the exposed cobbles out of the other side of the hole. Then when he is finished, you can switch sides. There will be a lot of free material to be sucked up, and far less cobbles to worry about for the nozzle-man; while there will be plenty of cobbles to move for the rock-man. As long as it can be done without clouding out the entire hole, this system will usually increase production.

It works well to place the rock-man in charge of safety in the hole. While the nozzle-man's attention is on the gold, in the material, and on the bedrock, the rock-man's attention is already onto moving the boulders and cobbles out of the way anyway. So he should have the responsibility to make sure that no potentially dangerous ones are hanging around waiting to fall into the hole and hurt someone.

I have also found that if the nozzle-man is responsible for freeing up his own plug-ups, then he will be more careful about what he sucks up into the nozzle, and he'll get fewer plug-ups! When I get a plug-up, I'll hand the nozzle to the rock-man so he can prevent further material from being sucked into the hose until the obstruction has been freed up. Meanwhile, he continues to move cobbles out of the hole with at least one hand.

The thing about suction hose, is if nothing but clear water is being run through it, you cannot always tell if an obstruction has been totally freed up just by looking at the suction hose. If the water that you are dredging in is not too clear, or if the hose is not very clear, or if the water is deep or fast, sometimes you can spend a lot of time looking for a plug-up and not know that it has already freed up on its own. Sometimes you will get a plug-up and go look for it; and in not being able to find it, you'll go to the surface to knock it out of the jet with the plugger pole—only to get there and find out that the obstruction already freed up minutes ago. Or sometimes you will think that it did, only to go back down to the nozzle and find out that you still have a plug-up somewhere in the system...

This all adds up to a lot of lost production time and wasted energy output, which can be avoided by having the rock-man suck up material as soon as he feels the obstruction free up. (He just needs to be extra careful to not cause a new plug-up, or he'll have the nozzle-man mad at him!) In this way, while the nozzle-man is looking for the plug-up, if he suddenly sees material moving up through the hose, he'll know that the obstruction has been freed up. And if he does not see any material moving up through the hose, he'll know he still needs to find the plug-up.

WATCH OUT FOR THE CON ARTISTS!

There are three areas in gold dredging where you are most likely to run into people who are going to try and pull the wool over your eyes, while they steal the money and gold out of your pockets. These are: mining claim sales, used mining equipment sales, and when you are selling your gold.

BUYING CLAIMS: Let's take up buying a claim first. They sell for just about any amount—depending on how much value is being attached to them. Notice I didn't say: "How valuable they really are!" Mining claims can sell for $200, and sometimes they are advertised for as much as a million dollars!

The best way to determine the *ACTUAL* value that a claim holds for you, is to thoroughly sample it. If you are considering to buy a claim that you intend to dredge, it is not good enough to take samples up on the banks with a gold pan. It really takes going out there with your equipment and dredging lots of holes, if you want a somewhat accurate index of how valuable the claim is going to be to you as a gold dredger. And further, it is not good enough to dredge just one or two sample holes and stop there if you find pay dirt. If you are going to pay a lot of money for the claim, it's best to dredge more holes to see how long and wide the paystreak actually is. Sometimes you can choose just the right spot to dredge a sample hole and land in beautiful pay dirt, only to find out later that it was a single deposit, or a very small stringer! This can be very misleading, and I've seen it happen a few times.

If a claim owner is asking a lot of money for a mining claim, based on the fact that it is "very rich," then he should have no objections to you doing some testing to see for yourself, right? When you see objections to this, you know there must be something wrong.

What you have to watch out for, is the claim owner who plays it up real big and gets you so excited about all the gold you're going to recover, that you do not feel the need to do any testing! "Oh man, this claim is so hot that it kills me to sell it–but I'm too old to work it and I need the money . . . " Here is where the con-man is going to try and get you.

When dealing with mining claims, if it is worth a lot of money–then it is definitely worth a lot of sampling first!

Another way that you can be stung on a claim deal is by buying one from someone who does not really own it–or by paying royalties to someone who has given you permission to dredge on the claim, but who doesn't really own it. It is not very difficult to locate a claim in the County Recorder's Office, and find out who really owns it–if anyone, and then to contact the BLM to make sure the yearly paper work has been kept up. Perhaps this isn't necessary when you first make a percentage deal over a claim; but if you get into large amounts of gold, it could be worth the little time it takes to verify who the actual owner is.

If you are going to buy a mining claim, you will always want to verify actual ownership in this way; and if you are going to put a lot of money up for a claim, it may even be worthwhile to have a competent attorney check into it also.

BUYING USED DREDGES: Before you buy a used dredge, you really ought to take it out onto the river and run it to make sure that it operates properly. This might not be positively necessary when it is a used standard store bought type of dredge; but it's definitely a must when you are considering buying a home built dredging rig which you haven't operated yourself yet. There can be many bugs that have yet to be worked out of a home built dredge which might not be apparent until you take the dredge out and run it—especially in pay dirt if possible. This is really the only way to see what you are buying.

SELLING YOUR GOLD: There are plenty of legitimate gold buyers around who will give you a good deal on your gold. There are also plenty of gold buyers around who would prefer to give you a lousy deal on your gold. Your best bet is to have a number of different buyers available to sell your gold to. When you are ready, you make some phone calls, and sell to whoever is willing to offer you the most.

If you are short on gold buyers within your own general operating area, drop *MINER'S REFERRAL* a line; they will probably be able to set you up with more. For information about *MINER'S REFERRAL,* see page 158.

Chapter 5

ADVANCED RECOVERY SYSTEMS

The smaller that a particle of gold is, the more difficult it will be to recover it in a sluice box. It's been estimated that a 1/4-grain sized particle of gold can be divided into at least 2000 separate particles of gold--each recognizable as a single color. There is an abundance of gold in any gold bearing river which is so fine in size that it is almost completely dissolved into a state of suspension in the river water itself. You are not likely to trap a substantial amount of gold that fine in size in a conventional sluice box; because it is so small and light that it will flow right along with the water wherever it goes.

I know a very experienced old-timer who has excellent fine gold recovery on his dredge. One time when he was dredging up a paystreak, for an experiment, he had a buddy hold up a 2 1/2-foot square amalgamated plate just behind his dredge where the silt would pass over it while he pumped material out of the paystreak. He says that it only took about two minutes for the plate to turn yellow from the ultra fine gold which passed over and out of his sluice box! The gold was extremely fine—so fine that there was only a small trace of it showing in his sluice box concentrates.

Any sluice box will be found to effectively recover gold down to a specific size of particle classification. We can call this the *"effective size range of gold recovery"* A sluice will start losing more particles of gold when they are smaller in size than the effective size range. The smaller the particles, then the greater percentage of them will be lost.

No gold dredger likes to be told he is losing any gold out of his sluice box—especially when some other dredger is standing there ridiculing his recovery system, telling him how much gold he is wasting, etc. So you run into a lot of miners out in the field who "know" their own recovery systems are the best thing going. This is ridiculous really; because anyone who's been around at all, ought to know by now that MY dredge's recovery system is the best!!

Actually, it is better to know that the real test is how fine in size your recovery system will still recover a large percentage of the gold particles.

There is a cutoff point for the effective range of particle size recovery; and if you are seeking to improve your recovery, it will be necessary to work on improving the system's efficiency in catching more of the smaller sized particles.

A dredger can pan off a set of his concentrates, see some ultra fines in his pan, and get the mistaken idea that his dredge is recovering effectively down to that particle size. Any concentrating type of recovery system would be very suspect if it were not catching at least some of the ultra fines. The point here is that you cannot see the fines and ultra fines which were washed right through and out of the sluice box. You can see, however, that you are recovering gold in an abundance down to a certain size range, and this will give you some indication of what your effective range of recovery is. To see some small amount of ultra fine sized particles in your concentrates, just proves that they are there in the streambed material. The key question is: How much is there?

The purity of gold also has a lot to do with how small in particle and flake size that your recovery system will be able to effectively recover it. The less pure that a specific particle of gold is, the lower its specific gravity will be; and so the slower the water force will have to be through the sluice in order to recover it. As an example, your recovery system might recover 90% of the 100 mesh sized gold particles when it has a purity of 88% gold. The same recovery system—unchanged in any way—might only recover 10% of the 100 mesh sized particles which are only 78% pure. So when you move from one location to the next, if the purity of the gold in the new area has changed, the efficiency of your sluice box's fine gold recovery might appear to change also.

FINE GOLD RECOVERY

To recover large percentages of fine and ultra fine sized gold, the water force through the sluice box needs to be slowed down. To do this, the larger sized materials have to be classified out and separated; because it takes a far greater force of water to move them through the recovery system. If you can separate out the larger sized rocks and materials, and direct the smaller sized materials and gold over a slow moving recovery system, your effective range of recovery will drop down to include smaller sized gold particles; and so your recovery can be significantly improved.

LARGER SIZED DREDGES

The larger the size of a dredge, the more that classification is needed; because more material and larger sized rocks are being moved—which takes a greater amount of water force. Also, larger sized rocks will cause more water turbulence around, behind, and below them as they are being washed through a sluice box. This additional turbulence can boil fine sized gold out of the riffles.

In addition, a larger sized dredge is able to process a much greater volume of streambed material. So when using one, if you can recover the fines and ultra fines, you can make lower-grade material pay off well. If you can make lower-grade material pay off, there will be far more acceptable pay dirt available for you to dredge in most of the gold bearing rivers; and so you can produce more gold, more of the time.

The way in which smaller materials are being separated from the larger sized materials in hydraulic type recovery systems (sluice boxes), is by washing the materials over top of a classification screen or punch plate, and directing the classified materials out from underneath the screen and into a slow moving recovery system. This is primarily being done today in three different ways: Triple Sluice Boxes, Double Sluice Boxes, and the Water-dam Classification System.

TRIPLE SLUICE RECOVERY SYSTEM

The triple sluice has a classification screen at the head of the box, which allows a lot of the smaller sized materials to be separated out and directed into side boxes, where the flow of water and material is controlled to move much slower. Most flakes and particles of gold are of a size which will drop through the classification screen and end up in the slower moving currents of the side sluice boxes. (See Figure 5-1.)

Because the triple sluice is still a hydraulic type of recovery system (water force–hydraulics–is used to wash the larger sized materials over the classification screen and through the sluice box), some of the fines and ultra fines are going to be washed into the center box also. The center box will recover fine gold as well or better than a conventional single sluice box; because it contains the same kind of riffle system, only not nearly as much volume of heavy concentrates will be washed into the center box; so the riffle system is not going to load up as heavily with black sands as it would in a single sluice box.

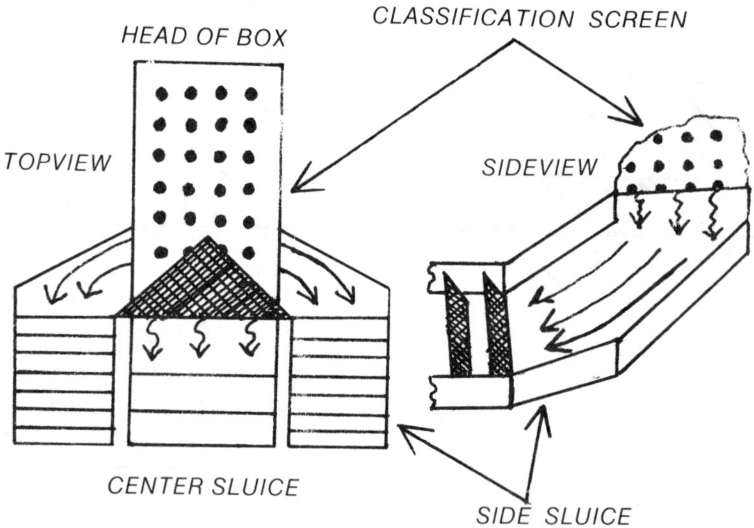

Figure 5-1: How a triple sluice system works.

The main advantage to a triple sluice is that a large percentage of fines will be washed into the side sluices, where they will not be thereafter affected by a fast turbulent current again. So the effective recovery of smaller sized gold flakes and particles is improved.

The best bet with a triple sluice (and a double sluice), is to try and put as much of the smaller sized materials down through the classification screen as possible. This way more of the fine gold can be directed away from the main fast stream of water and larger sized materials. The length of the classification screen, before it reaches the side sluice box openings, can make a big difference as to how much classified material is washed into the side boxes. The longer that the classification screen is, the more classified material can be washed down through the screen.

Triple sluice boxes can be beneficially modified by extending the length of the sluice box and classification

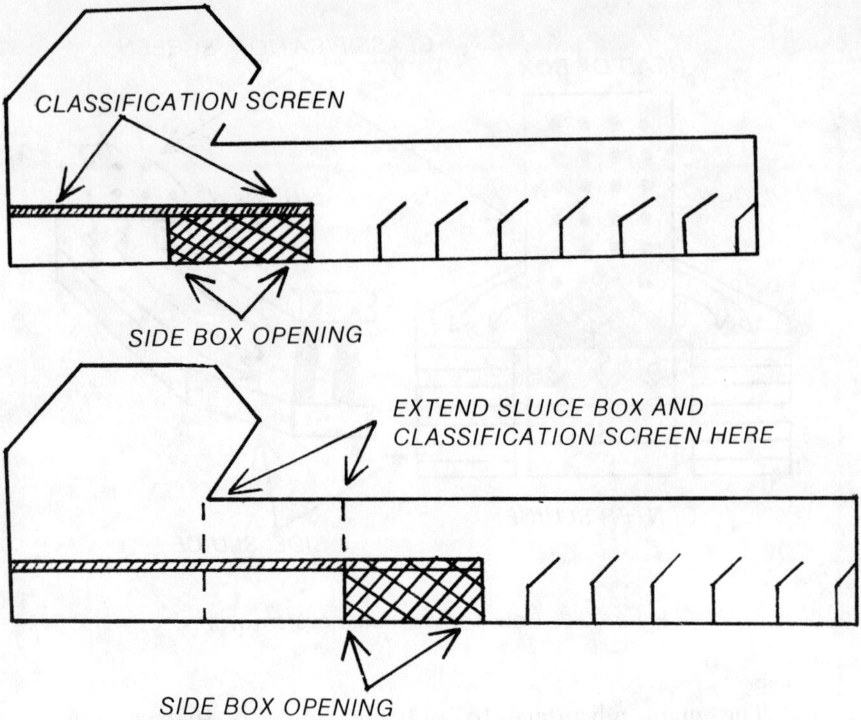

Figure 5-2: Triple sluice modification allows more fine gold values to get into the side sluice boxes.

screen in front of the side box openings, as shown in figure 5-2. This will increase the percentage of fines and ultra fines that are washed into the side sluice boxes.

Keep in mind, however, that this modification is also likely to require additional flotation underneath the sluice box at the rear-end of the dredge.

Ideally, with a triple sluice recovery system, most or all of the smaller sized materials will be washed down through the classification screen and into the side boxes. You have to watch this though; because when you are pumping sand or pea gravel, more materials can be washed into the side boxes than the slow currents there can wash out of them; they can load up with material; and you can sink your dredge! So when building or modifying a triple sluice dredge in this way, it's best to devise some sliding doors across the side box openings so you can regulate the amount of current which flows into the side boxes—or shut them off altogether when you are going to be pumping large amounts of valueless sand or pea gravel.

DOUBLE SLUICE RECOVERY SYSTEM

One of the latest developments on the gold mining equipment market is a kit to convert the old conventional single sluice boxes into improved, fine gold recovering, double sluice boxes. It is done by classifying out the smaller materials with a screen at the head of the sluice box—just like in a triple sluice system. Then the classified materials and gold are directed into a separate slower moving sluice box located right underneath the main flowing upper standard sluice. (See figure 5-3.)

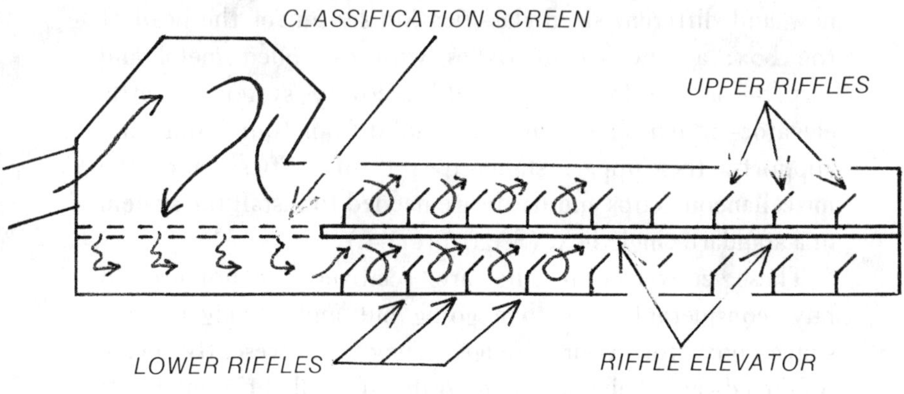

Figure 5-3: A single sluice can be converted into a double sluice to improve gold recovery.

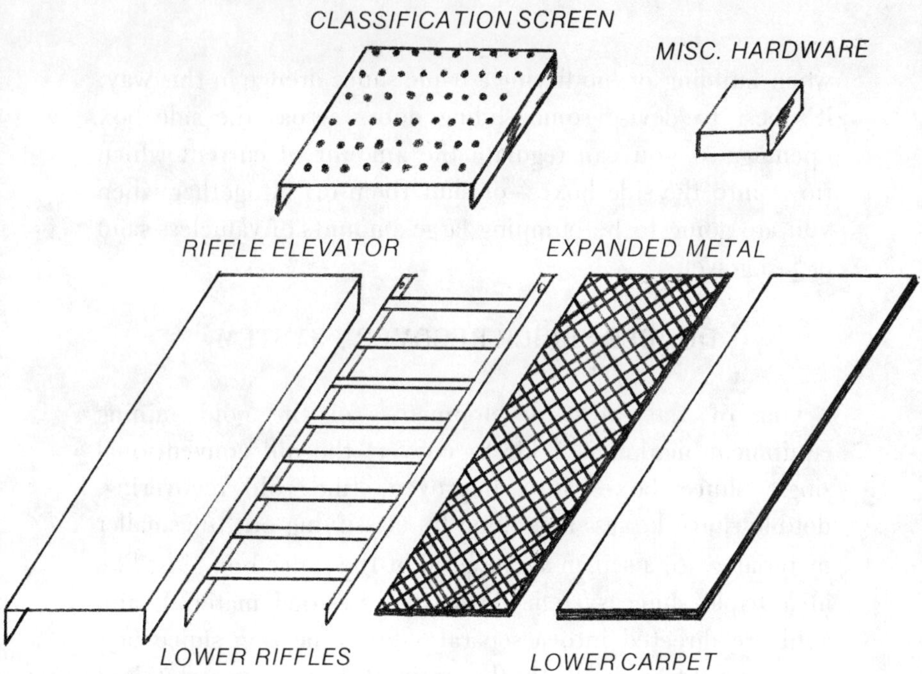

Figure 5-4: Double sluice conversion kit.

The double sluice conversion kit is simple; it includes a new and different sized classification screen for the head of the box; a new set of riffles with expanded metal and carpet–for the lower fine gold recovery system; a riffle elevator–which simply acts as a solid floor foundation that supports the upper standard set of riffles; and the miscellaneous clips and hardware needed to install the system in a standard single box. (See figure 5-4.)

These conversation kits are relatively economical to buy–considerably less than going out and buying a triple sluice box for your dredge. They are presently being produced–and being made to order–for all different kinds and sizes of single sluice boxes by: *KEENE INDUSTRIES,* 9330 Corbin Ave., Northridge, CA 91324, (213) 993-0411. Be sure and carefully measure the exact inside dimensions of your sluice box before you order.

WATER DAM CLASSIFICATION SYSTEM

There is a third way being used out in the field to effectively separate the smaller classification of materials from the larger, and place them in a slower moving current. It is being done by installing a classification screen (expanded metal) over top of the riffles in a single sluice recovery system to keep the larger sized materials out of the riffles, as shown in figure 5-5.

The classification screen must not be placed too far above the riffles in the box; otherwise there will not be enough water force above the screen to wash the larger materials out of the sluice.

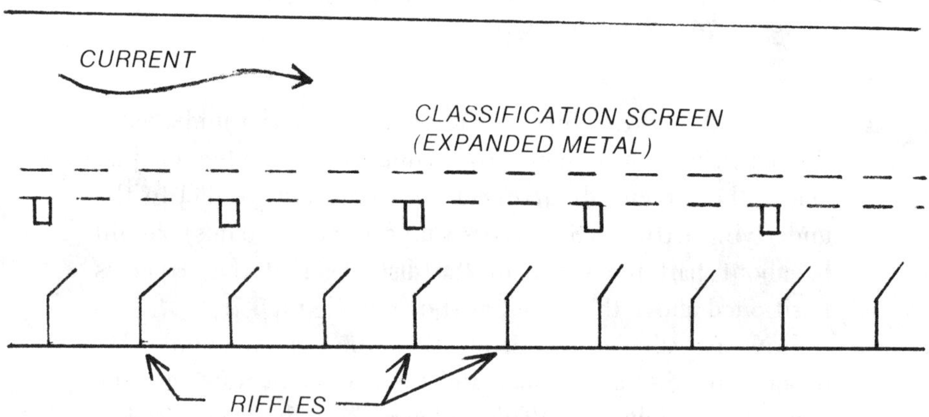

Figure 5-5: A classification screen can be used to keep the larger sized materials out of the riffles.

The turned up type of expanded metal works well with this kind of system—with the turned up points pointing *WITH* the current in the box, as shown in figure 5-6. In this way, they will not jam with pebbles, and they also will not direct the fast upper current down beneath the screen into the riffles.

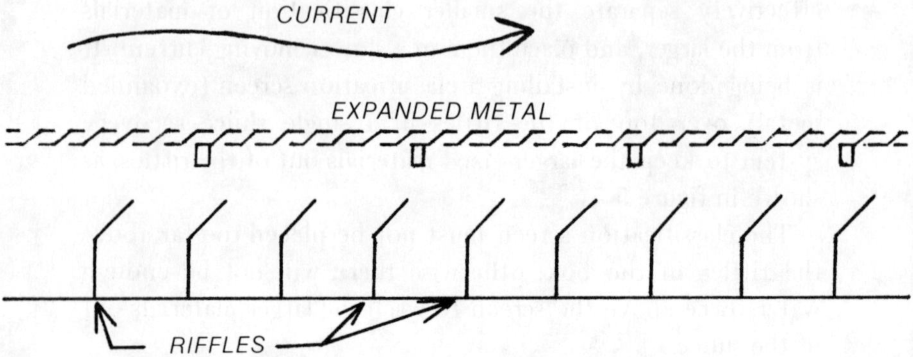

Figure 5-6: When turned up expanded metal is used as a classification screen, the turned up points should be pointing WITH the water's flow.

To further slow down the water and material underneath the classification screen, cross supports are extended just below the screen, about one foot apart, and parallel to the underlying riffles. These cross supports (water-dams) should be about half the width of the distance that the screen is positioned above the riffles, as shown in figure 5-7.

How far above the riffles that the screen should be, depends on the dredge size and the volume of water moving through the sluice box. This will vary from dredge to dredge. For example, in a five inch sized dredge, if the screen were two inches up above the riffles, then the water dams should be about an inch wide.

The reason why they are called "*water-dams*" is because they do just that; they are barriers which slow down the water and classified materials moving underneath the screen just over the riffles; so they improve fine gold recovery. They also create a low/back pressure area just behind them, which tends to pull additional classified material through the screen where they can be washed into the slower moving current over the riffles. (See figure 5-7.)

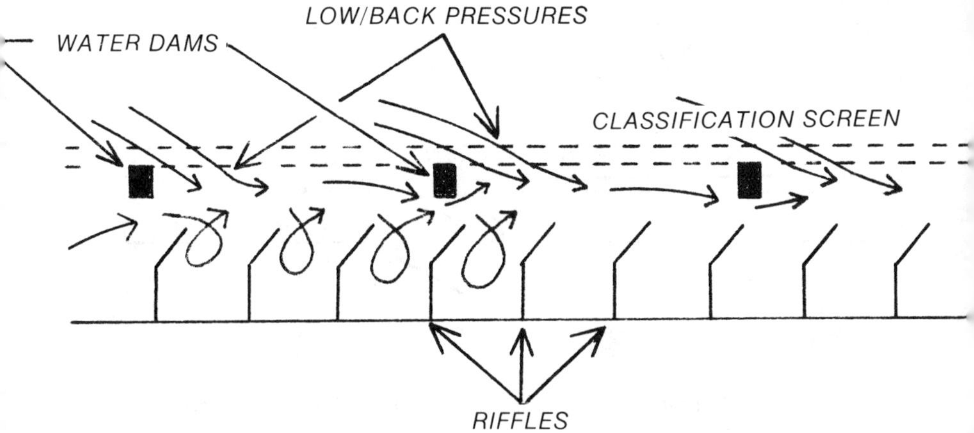

Figure 5-7: Water Dams slow down the current below the classification screen, and set up a low/back pressure area just behind them, and pull more material down below the screen.

The main trick to this water-dam classification system is in placing the screen close enough to the riffles, so there will still be enough water force to push the larger sized materials through and out of the sluice box; but not so close that the smaller sized materials are not given plenty of opportunity to work their way through the screen and into the riffles. The proper height of the classification screen is going to vary depending on each individual dredge, and some experimentation will be necessary to get it perfect; but when it's right, fine gold recovery can be greatly improved over that of the standard single sluice box type of systems—especially when larger amounts of volume are being processed through a larger sized dredge.

When a water-dam classification system is being installed into a single sluice box, it is a good idea to leave at least three or four riffles in the box that are not screened off. These will act as traps for any of the larger sized pieces of gold which are too large to pass through the classification screen. (See figure 5-8.)

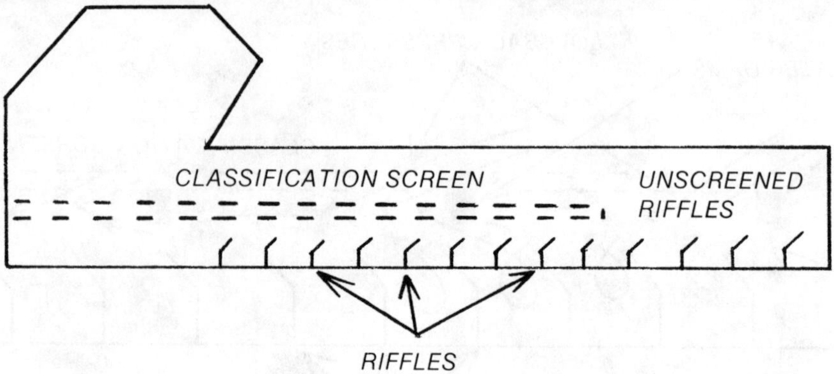

Figure 5-8: Leave 3 or 4 riffles open to trap the larger sized nuggets too.

A FEW OTHER POINTS ON RECOVERY . . .

SUCTION HOSE AND GOLD: As material is being sucked up through a length of suction hose, just like when it is being washed down along a riverbed, the gold will quickly move through the material to the bottom of the suction hose, where the water and material will be moving slower because of the friction barrier. So the majority of gold will be moving slower, and will be riding up along the bottom of the suction hose as it enters the jet.

DUEL EDUCTORS AND GOLD: On larger sized gold dredges, the dual eductor power jet is very efficient for several reasons; first, there will be a more orderly mixing and less of a friction power loss inside the main jet tube, as compared to the single eductor system. (See figure 5-9.)

The dual eductor power jet system is efficient for another reason too. If the jet eductors are placed horizontally, as the water force enters the main jet tube, it will not greatly disturb the gold which is riding along at the bottom of the hose and jet when it reaches that point. This way, the

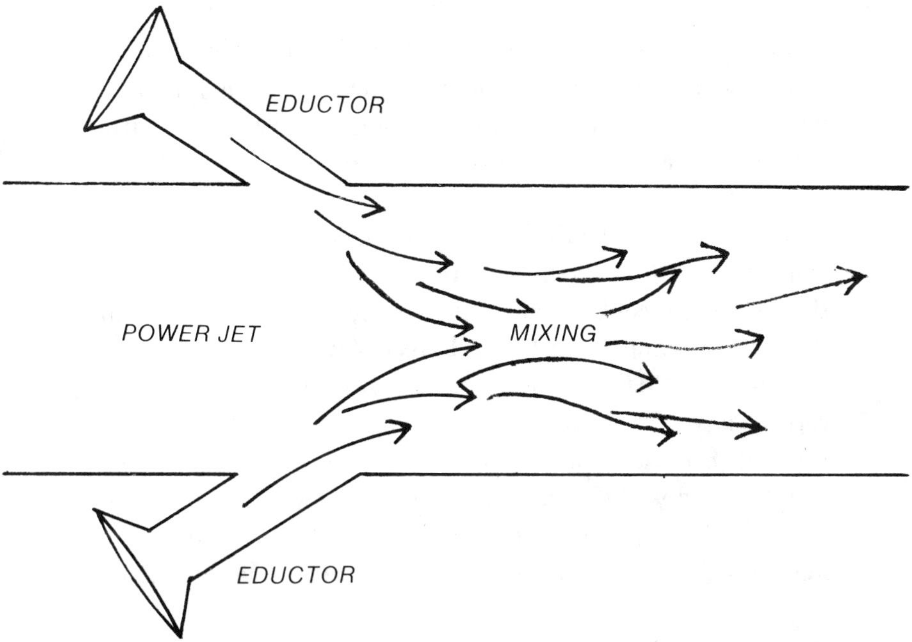

Figure 5-9: Duel eductors on a power jet will provide better mixing and less of a friction power loss.

material is allowed to enter the sluice box with the gold still riding right along at the bottom where you want it—and where 20-25 feet of travel has placed it for you. Single eductors, triple eductors, five eductors, and couple jet systems do not allow for this, and will blow the gold up off the bottom of the jet and into the material again just before it enters the sluice box.

The proper angle of an eductor as it enters the main jet tube is about eleven degrees. This will provide optimum mixing without greatly disturbing the gold along the bottom of the jet.

FLARED JETS AND GOLD: Flared jets are a good idea; they slow the water and material down before it enters the box, and they also allow more of the fine gold to drop to the bottom of the material before it enters the box. Most flared jets are flared out on a two to one ratio; an eight inch jet would be 16 inches wide where it enters a sluice box. I ran into a guy last summer who was using a three to one ratio, and he was having excellent results—although this may be too wide for the triple sluice systems.

ALASKAN DAMPERS: Long and heavy Alaskan dampers work better than short ones to slow down the water flow through a sluice box, and seem to be more effective than short ones at improving fine gold recovery.

NO-MAD CARPET: Keep your eyes open for "No-mad" carpet; it works great underneath your riffles. It's the same type of material that many department stores lay out for people to wipe their feet on. Some carpet stores sell it to lay underneath carpet to act as a soft padding cushion. Most miners call it "plastic spaghetti matting," because that's

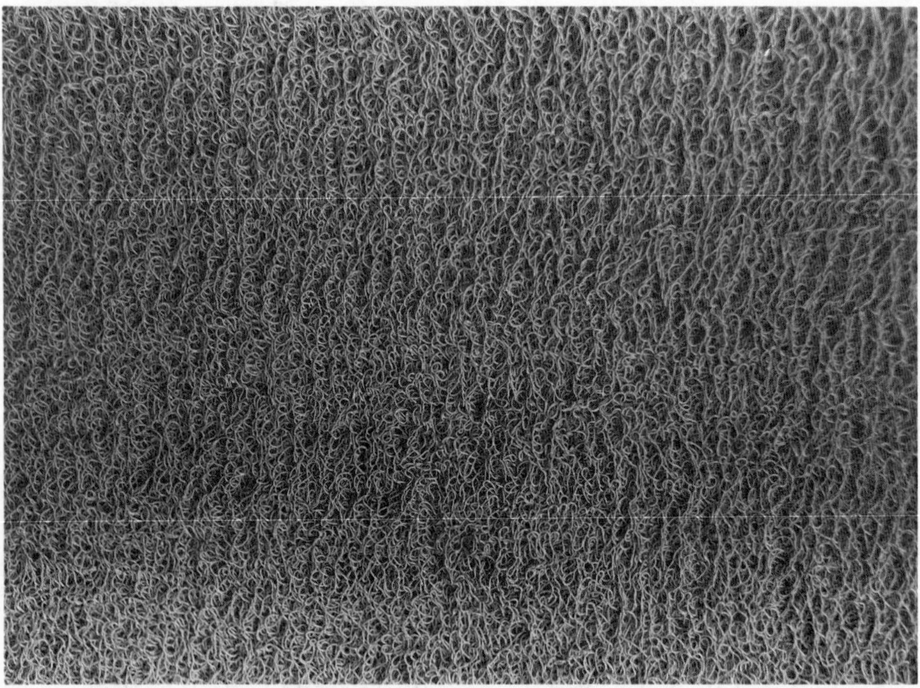

Figure 5-10: "No-mad" type carpet.

exactly what it looks like! Some of the dredge shops are now selling it. (See figure 5-10.)

No-mad carpet comes with and without a backing to it—it works great either way. A length of carpet or burlap should be placed underneath of the unbacked type if it's used in a sluice box.

ARMOUR WEAVE RIFFLES: Also keep your eyes open for the armour weave type of expanded metal. This is a taller kind of turned-up expanded metal, which acts as a great recovery system in the side boxes of a triple sluice, or in a single sluice box underneath a water-dam classification system. Armour weave is also lighter than regular expanded metal. It comes either in aluminum or in steel. Sometimes it is hard to find; you'll have to call around to the various steel and metal dealers.

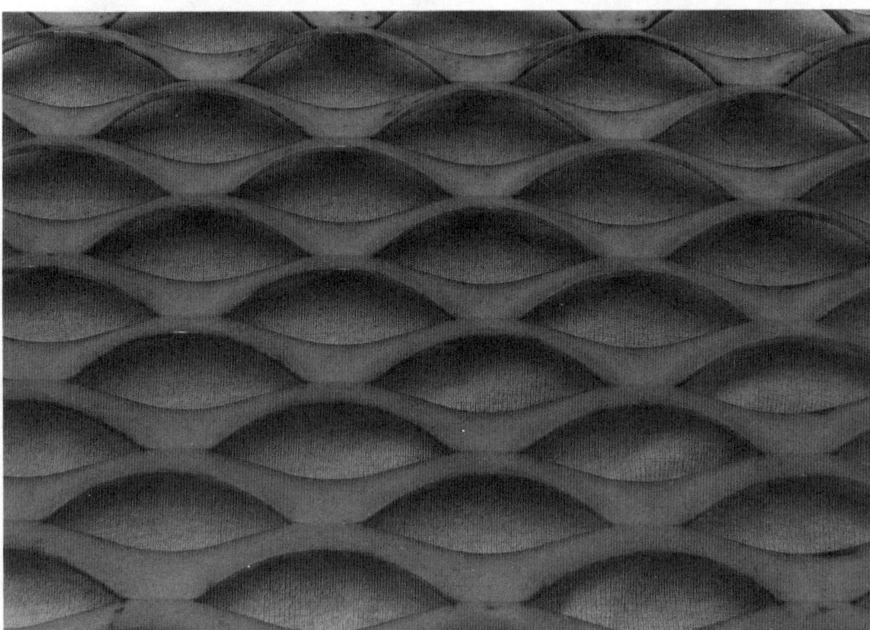

Figure 5-11: "Armor weave" type expanded metal makes a good riffle system.

MINER'S REFERRAL

Miner's Referral is a service with the purpose of bridging the communication gaps amongst individuals and groups within the mining profession.

The three areas which it is presently working very hard at, are to:

(1) Connect up miners with good proven gold locations–and with the claim owners who want to lease out their claims, or allow miners to work them on a percentage basis or otherwise.

(2) Connect up miners with the gold buyers in their general areas so that miners can get the best prices for their gold.

(3) Accumulate the names and a means to contact miners, and mining related groups, so that the entire mining community can be contacted and better organized in any time of need when our mining rights are threatened. All miners, mining groups, and others who are interested should forward a writing address to the below address–and keep it up to date–so they can be on the receiving end of any vital news which may concern them.

If you buy gold, or if you own mining claims and want others to mine on them on a percentage basis, lease, or otherwise; or if you are a claim broker, claim finder, or have claims to sell; please forward your name and address to the below address and let them know what counties the claims are within. Miner's Referral will pass on this information to those miners who are interested in mining in your general area so they can contact you.

If you want to receive data about promising locations or the open claims within a specific location—or the gold buyers there, send along a self-addressed stamped envelope, and two dollars for each specific referral that you ask about. In exchange, Miner's Referral will send you what information they have accumulated—if any—about the specific areas of your interest.

MINER'S REFERRAL
P.O. Box 1932 Weaverville
CA 96093

AUTHOR'S NOTE

Feel free to write the author.

I am very interested in hearing from you regarding this book, mining in general, or about anything else which you would like to communicate.

All mail sent to the below address will be received by me, and answered by me directly.

Dave McCracken
P.O. Box 702
Big Bar, CA 96010

GLOSSARY

ACCEPTABLE REQUIREMENT: The minimal amount of gold that you need to recover on an average daily basis to make a certain area worth dredging on a production basis.

ACCESSIBILITY: 1) Having legal rights to bring your equipment into a location and mine it. 2) Being able to get your equipment into a location without a great deal of trouble.

ALASKAN DAMPER: A device which attaches to a sluice box, usually made of rubber, which rides along the surface of the water as it moves over the box. The "damper" does various things which improve the recovery of the sluice box, including smoothing out the flow of water over the box, and breaking the surface tension of the water.

ANCIENT MATERIALS: Are streambed materials that are either located over virgin ground, or are materials which have been swept away from virgin ground during a major storm to be redeposited elsewhere.

ANCIENT STREAMBED: For our purposes would be natural streambed which has been in place since before the oldtimers arrived on the scene. Those gravels, and that area, have never been mined before.

ASSAY: An analysis of an ore sample, streambed sample, or a gold sample to determine the proportions of gold, silver, platinum, and other valuable metals, to the amount of waste material present in the sample.

BACK PRESSURE: When you see water moving backwards against the main flow of water--as in a back eddy, you are looking at a back pressure area.

BACK-TRACKING: Following the gold back through the shallow material of a paystreak in an effort to find the tail end. See figure 3-1 in Chapter 3 for a visual example.

BAILING OUT: To drop the weights which hold a diver underwater in order to quickly get to the water's surface.

BEDROCK STRINGER: A paystreak which is lying right on the bedrock.

BENCHES: Also known as "terrace placers." Sections of old streambed which have been left high and dry by the present stream of water are referred to as "benches." Many benches still contain large paying quantities of gold.

BLEEDING OFF: Natural erosion of bench gravels into a present stream of water.

BOIL-OUT: If too much force of water is moved over top of a back pressure area, the force of the flow being pulled into the vacuum becomes so turbulent that even most of the heavier materials--gold included--will be "boiled out."

BONANZA: A very rich gold deposit--whether it be a lode or placer--is often referred to as a "bonanza."

BOULDER: Rocks which are of a size larger than a large cobblet. For our purpose, a boulder is a rock which is large enough that you have to use your body weight in order to move it.

BOULDER BOUND: A hole which has so many boulders at the bottom that it is almost impossible to get a boulder harness around any of them because they are packed so close together.

BOWLINE KNOT: An easy to tie and untie sailor's knot, which has a nonslipping loop. There are many uses for this type of knot in a dredging operation. Tying instructions shown in figure 11-1 in Chapter 11.

CLASSIFICATION: The process of screening out the larger sized materials from a body of ore, streambed gravels, or a set of concentrates. Classification is usually done with the use of one or more sizes of mesh screening.

CLASSIFIED MATERIALS: Those materials which have been screened through a certain size of classification screen.

CLASSIFIER: A screening device itself is often referred to as a "classifier." There is a classifier commonly found at the head of most sluice boxes on gold dredges today.

CLASSIFY: See CLASSIFICATION.

CLEAN-UP: The action or removing the gold from a set of concentrates and then cleaning the gold so it takes on its natural look.

CLEAN-UP GEAR: Any equipment which is used in the process of clean-up, such as mesh boxes, gold pan, etc.

COBBLETS: Rocks which are too large to go through the suction hose intake nozzle, yet smaller than a boulder. Cobblets need to be removed from a hole by hand.

COMMON GOLD PATH: Because gold is six or seven times heavier than the average of other streambed materials that get swept downriver during a major storm, it tends to follow and be deposited along a common path. For the most part, this path seems to follow to the inside of the bends and the shortest route between the bends, as shown in figure 1-5 in Chapter 1.

CREVICE: A split, crack, or open fissure in the bedrock's surface. Those crevices that are located at the bottom of a streambed are likely to trap gold out of the material as it is passed over during a large storm.

DIRECTIONAL CHANGE BLOCK: (Winching) Is a heavy duty pulley used in winching boulders, which is anchored in a specific location so that boulders can be pulled in the desired direction. Examples shown in figures 6-6, 6-8, and 6-14 in Chapter 6.

DOUBLE BLOCKING: (Winching boulders) Is done by attaching a snatch block to the boulder sling, running the pull cable through that block (pulley), and then back to the last directional change anchor. This type of set-up will double the amount of pulling force against the boulder that is being exerted by the winch and pull cable. Example shown in figure 6-15 in Chapter 6.

DUMPING OFF: The act of removing the concentrates from a sluice box.

DUST: Extremely fine particles of gold are referred to as "gold dust." A good recovery system being used in some areas will recover a considerable quantity of "dust"--often enough to pay all of the operating expenses, sometime more. In the earlier days of mining, a pinch of gold dust was worth a dollar. Today, the same pinch is worth about 20-25 dollars.

EFFECTIVE SIZE RANGE OF GOLD RECOVERY: Any sluice box will be found to effectively recover gold down to a specific size of particle classification, and will start losing more particles of gold when they are smaller in size than this effective size range.

FAST WATER: When dredging in water which is swift enough that it will wash your body down stream if you are not extra careful to avoid it, it is considered to be fast water.

FAST WATER STABILIZERS: Extra floatation attached securely to each side of a dredge in order to increase its stability when moving the dredge out into fast water. Fast water stabilizers prevent your dredge from being capsized or swamped by the swifter moving water.

FILLER MATERIAL: Material which works its way down between the major rocks of a streambed as it forms. Filler material usually tends to harden over a period of time and cement the rocks together to a degree.

FINDER'S FEE: A percentage of the recovery given to a person because he helped to locate a paying deposit for an operation.

FINE GOLD: Those particles of gold that are small enough to pass through 40 mesh screen are known as "fine gold."

FINENESS: A system which is used to indicate the purity of a gold sample or specimen. A specimen having a fineness of .900 would be 90% gold; a fineness of .650 would be equivalent to 65% gold, and so on.

FLAT ROCKS: It is not uncommon to find an increase in the percentage of flat rocks in a low/back pressure area.

FLOOD LAYERS: Sometimes a streambed will have different layers of material that were laid down during different storms at different periods of time. Sometimes each separate layer contains its own gold in varying amounts. These different layers are called "Flood Layers."

FLOUR GOLD: Same as GOLD DUST. (See DUST.)

FLOURED MERCURY: Mercury which has been broken down into thousands of tiny bubbles.

FORWARD-TRACKING: Dredging a narrow path forward at the upper end of a paystreak along the line of where it paid the best, in an effort to find an extension of the paystreak if it is there. Example shown in figure 3-9 in Chapter 3.

FREE GOLD: Those particles of gold in an ore or in a set of concentrates which are not chemically locked in with the other elements.

FULL FACE MASK: A full sized mask which covers the entire face that is often used in cold water diving to prevent the face from being exposed directly to the cold.

GENERAL LOCATION: A general area in which a miner chooses to conduct his mining operation.

GRAIN: A term used to label a small particle of gold, and also as a unit of weight in the troy system of measurement--in which 24 grains equals 1 pennyweight. One grain also equals 64.8 milligrams.

GRUBSTAKE: The practice of a storekeeper extending credit--or another person putting up money for the purpose of outfitting a prospector and keeping him supplied until he makes a strike. In exchange, that person then receives an agreed upon percentage of the find.

GUT: The river's channel is often referred to as the "gut."

HARD PAN: The bottom of a flood layer and top of the layer just below is called a hard pan. The layer just below the flood layer is usually hardpacked enough that the flood does not wash it away too. The hard pan often contains the flood gold which was riding at the bottom of the flood layer.

HEADER BOX: The forward most section of the sluicebox upon most suction gold dredges today is designed to slow the materials down and spread them out evenly over a classifier as they enter. This upper section is referred to as the "Header Box."

HEMATITE: An iron ore ($Fe_2 O_3$) which is of reddish brown color. Hematite is one of the main non-magnetic minerals that comprises black sand concentrates.

HIGH BENCH: See BENCHES.

HIGH-GRADE MATERIAL: The richest portions of streambed contained in a paystreak.

HIGH-GRADING: Is when a person is stealing the higher-grade gold specimens, or a portion of the gold recovery, from a mining operation.

HOOKA AIR SYSTEM: Air breathing system used on most gold dredges in which the air is pumped from the surface down to the divers through an extended air line.

HOT WATER SUIT: Water can be heated up at the surface and pumped down through a hose to pour into a protective suit. In this way even an old wet-suit can be made to keep a diver comfortable when dredging in cold water. Full details covered in Chapter 8.

HYDRAULIC CLASSIFICATION SYSTEM: A means of material classification in which water (hydraulics) is used to wash the materials over top of classification screens. Double and triple sluices are examples of a hydraulic classification system.

HYDRAULIC MINING: (Hydraulicing) A method of above water gold mining in which a large volume of water is directed at gold bearing streambed, so its materials can be washed down through sluice boxes, where the gold can be recovered.

INTERMEDIATE: A term used to identify the medium sized dredges. Intermediates are usually of four or five inch suction hose sizes. Intermediates are light enough so they can be moved around without too much trouble. Yet they can also move enough material to pay well when used in a richer paying location.

JET: Is that component on a suction dredge that water is pumped into, and which creates a suction up through the suction hose. The jet is designed so that streambed material can be pumped directly to the recovery system without having to be run through the water pump.

LAYERS: Over a period of time, different storms deposit new material on top of older material. These different deposits are called layers. They can usually be recognized by their different colors and degrees of hardness.

LOCKED IN VALUES: Valuable minerals (gold) which are chemically locked in with other elements within the black sands or an ore, requiring more involved methods of processing so the values can be released and recovered.

LODE: Veins that contain valuable minerals are referred to as "lodes." Gold, in hardrock form, (lode) is commonly associated with quartz veins that protrude through the general country rock which makes up the earth's crust. Lodes are the original source of placer gold deposits.

LOOSE PACK: See "LOOSE PACKED MATERIAL."

LOOSE PACKED MATERIAL: Material which is not cemented together, but is loosely sitting there. Often when punching a hole in loose packed material, the sides of the hole keep sliding in because there is nothing to hold the material up.

LOW-GRADE MATERIAL: Streambed material which contains valuable minerals (gold), which might not be rich enough to dredge--or which might be just acceptable.

LOW/BACK PRESSURE AREA: Any area in a river or stream of water in which the water force slows down or reverses direction due to some sudden change in the bedrock, some kind of obstruction, or some change in the direction of flow. Gold tends to concentrate in the low/back pressure areas which lie along the common gold path.

LOW PRESSURE AREA: Any place in a river or stream of water in which the water slows down or stops because of an obstruction, a change in bedrock, or a change in flow direction. Low pressure areas can be small, and they can be large--like a long stretch of river which runs wider than it did upstream.

MAGNETITE: Fe_3O_4 The heavy magnetic black sands (iron) that are found in the heavy concentrates which collect in a recovery system.

MAJOR STORM AND RUNOFF: It takes an incredible amount of water force to redeposit the streambed and gold in a river. Winter storms, in most cases, are not enough to do this. Once in a while, though, a storm or snow runoff of huge dimensions will occur and redeposit the various streams and rivers in the area. We call these "major storms and runoffs."

MATERIAL: "Material" is a term that is used to indicate all of the rocks, sand, gravel, mud, clay, and silt which makes up a streambed.

MERCURY: ("Quick silver") A heavy, silvery colored, liquid type of metal which has a tremendous affinity for gold, silver and many other metals. Mercury is often used amongst gold miners to collect the fine gold values out of a set of heavy concentrates that have been taken from a recovery system. The procedure is called "amalgamation."

MESH: Signifies the number of openings that are contained in a lineal inch of screening. For example, 10 mesh screen will have 10 openings per lineal inch and 100 openings per square inch.

NITRIC ACID: A clear, fuming, highly corrosive liquid that is often used to clean gold of various impurities.

NOZZLE-MAN: The dredge operator who controls the suction nozzle is normally referred to as the "nozzle-man."

OBSTRUCTION: The rock or group of rocks which cause a plugup in a suction hose or jet.

OVERBURDEN: Is the lower-grade streambed material that lies on top of a placer paystreak. In dredging, the overburden must be removed first, before processing of the paystreak is possible.

PAY: How much gold that a lode or streambed is yielding to a mining activity is often expressed as how much it is "paying."

PAYSTREAK: A paystreak is a concentration of gold which has formed in a section of river or stream because of a lessening of water force in that area during a storm.

PENNYWEIGHT: A unit of the troy system of weight measurement. One pennyweight equals 24 grains, and is 1/20th. of a troy ounce.

PLACER DEPOSIT: Free gold that has eroded away from its original lode and which has been swept into a stream of running water will tend to accumulate in certain common locations. These accumulations are called "placer deposits."

PLATINUM: Is a family of 6 rare and valuable metals which are usually silvery white in color. Because platinum is sometimes recovered along with gold, it's well for the placer miner to know what it looks like so he does not discard it along with the waste materials from his recovery system.

PLAYED OUT: All gold deposits--whether placer or lode--will run out sooner or later. When all of the known about paying quantities of gold or other valuable minerals have been mined out of such a deposit, it is said to have "played out."

PLUGGER POLE: A long thin rod made of metal or PVC piping that is used to tap the plug-ups out of the jet on a dredge from the surface when they occur.

PLUG-UP: A plug-up is when a rock or combination of rocks lodge in the suction hose or jet and prevent the passage of material through the hose or jet.

POKE: A miner's container of gold. In the earlier days of mining a small leather bag was often used to hold a miner's collection (poke) of gold. Today, plastic and glass jars are most commonly used for this.

PRIMER: That device on a suction dredge which connects the water pump to the water, which is also designed to be manually filled with water so the pump can be easily primed from the surface.

PROSPECTING: The activity of attempting to find a deposit of gold or some other valuable mineral. Prospecting usually involves lots of moving around and taking numerous samples until an acceptable deposit is located--at which time mining activities can begin.

PUMPING MATERIAL: The action of feeding material into the suction nozzle. The material rides up through the suction hose and up into and over the sluice box where the gold is recovered from the material.

PUNCHING HOLES: See SAMPLING.

RECOVERY SYSTEM: That part of a mining operation or mining equipment which is designed to recover the gold from the material which you are running through the system. The recovery system on most dredges consists of a sluice box.

RICH: Streambed material which contains acceptable paying quantities of gold--or better, is often said to be "rich."

ROCK-MAN: When two men are working together on a dredge underwater, one man will control the nozzle (nozzle-man), and the other will be his helper and move rocks out of his way (rock-man); so the nozzle-man can suck material into the nozzle with fewer obstructions in his way, and increase production.

ROLLING HITCH: (Winching) A rolling hitch is done by slinging the boulder backwards, and then running the chain or pull cable over top of the boulder. This will usually free the boulder up by rolling it. Example shown in figure 6-12 in Chapter 6.

ROYALTIES: Royalties means payment. A claim owner usually receives a percentage of what an operation finds on his claim. A grubstaker may also receive a percentage. These payments are often referred to as "royalties."

SAMPLE HOLE: When you want to check out a location to see if it is rich enough to dredge for its pay value, you dredge a series of holes in the location in an effort to determine how much gold is there. These are called "sample holes."

SAMPLING: Basically consists of exploratory testing in an effort to locate acceptable paying ground prior to starting a production mining operation. This is usually done by dredging sample holes in the likely spots to find gold deposits, should they be present within that general area. The material from each sample hole is tested to check the gold content. The prospector continues to make sample holes until a sufficient paystreak or deposit is located, at which time mining activities are begun with the intention of recovering the gold out of the deposit. Full procedures covered in Chapter 2.

SCUBA: Self Contained Underwater Breathing Apparatus (scuba tank, etc.) was used during the earlier days of suction dredging,but has since yielded to the hooka air breathing systems--which are more efficient for most of today's dredging activities.

SILVER: A shiny metalic element that is usually found in close association with gold to some degree. Silver is also considered to be a precious metal and is superior to any other metal in its ability to conduct electricity and heat.

SINGLE DEPOSIT: A single deposit of gold can form anywhere along the general path in the river which gold follows, where an obstruction, a small change in bedrock, or a crevice is located.

SLOPE: Is a term which is commonly used to express the downward gradiant of a sluice box or streambed.

SNATCH BLOCKS: (Winching) A type of pulley which can easily be attached or unattached to a cable.

SPECIFIC GRAVITY: Is the ratio of the weight of a volume or mass of any substance, as compared to the weight of an equal volume of water. The specific gravity of gold is 19.3, which means that any sized mass of pure gold will be found to weigh 19.3 times as much as an equal volume of pure water.

STRINGER: Same as PAYSTREAK.

SWEPT BEDROCK: Bedrock which has too much water flow over it to allow any bed to form. It is clean bedrock. There is no material sitting on top of it.

TAIL END: The lower end of a paystreak where the streambed material ceases to pay in acceptable gold values.

TAILINGS: The material which rides across and off of the recovery system is know as tailings.

TAIL RACE: The channel where the tailings are washed through after they exit the recovery system in a surface mining operation.

TROY WEIGHT: Is a system of weight measurement which is commonly used amongst miners and gold dealers.

24 grains . . . . = 1 pennyweight
20 pennyweight . . . . = 1 troy ounce
12 troy ounces . . . . = 1 troy pound

ULTRA FINE GOLD: Is a generalized term referring to gold which is very fine in size. See DUST.

UPCROPPING: An extension of bedrock which extends up out of the foundation of a streambed. Upcroppings can be the cause of single deposits or paystreaks when they lie along the common gold path.

VALUES: Those minerals that are contained in streambed materials that are of value to the gold dredger.

VIRGIN GROUND: Is an area that has never been mined before.

VIRGIN STREAMBED: Any material which has been redeposited into a naturally formed streambed during a major storm could be classified as "virgin streambed" because it has yet to be mined since having been laid down--even if it has been redeposited into a location that has been mined before.

WASHTUB: Large container which is often used to wash the concentrates out of a sluice box into.

WATER DAM: A wide support which holds a classification screen up over a set of riffles, and which slows the water down below the screen so that fine gold recovery can be improved.